MERLIN'S
TOUR
of the
UNIVERSE

Neil de Grasse Tyson

DOUBLEDAY

New York

London

Toronto

Sydney

Auckland

MERLIN'S TOUR

of the

UNIVERSE

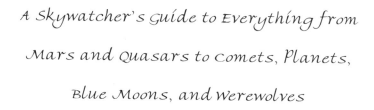

A Skywatcher's Guide to Everything from

Mars and Quasars to Comets, Planets,

Blue Moons, and Werewolves

A MAIN STREET BOOK
PUBLISHED BY DOUBLEDAY
a division of Bantam Doubleday Dell Publishing Group, Inc.
1540 Broadway, New York, New York 10036

MAIN STREET BOOKS, DOUBLEDAY, and the portrayal of a building
with a tree are trademarks of Doubleday,
a division of Bantam Doubleday Dell Publishing Group, Inc.

BOOK DESIGN BY CAROL MALCOLM RUSSO / SIGNET M DESIGN, INC.

Library of Congress Cataloging-in-Publication Data
Tyson, Neil de Grasse.
Merlin's tour of the universe : a skywatcher's guide to everything
from Mars and quasars to comets, planets, blue moons and
werewolves / Neil de Grasse Tyson. — 1st ed.
p.   cm
Includes bibliographical references and index.
1. Astronomy—Miscellanea.   I. Title
[QB52.T98   1997]
520—DC21   96-37200
CIP

ISBN 0-385-48835-1
Copyright © 1989, 1997 by Neil de Grasse Tyson

*T*o all people
who recognize herein
a question of their own

.

CONTENTS

CONTENTS

MERLIN'S TOUR OF THE UNIVERSE is a collection of questions asked by the general public and answered by Merlin, a visitor from the Andromeda galaxy who is as old as Earth and has observed the major scientific achievements of recorded Earth history.

Merlin writes a question-and-answer column for *Star Date,* the astronomy magazine published for the lay reader by the McDonald Observatory of the University of Texas at Austin. Most of the material in *Merlin's Tour of the Universe* first appeared in this column.

This book is not intended to be an astronomy tutorial. Its focus is guided entirely by questions asked by interested readers who have ranged from age four to age ninety. I have written these answers for the enjoyment of my readers and with my life's enthusiasm for this beautiful and grand home that is our universe.

$\mathcal{M}$ ERLIN WAS BORN nearly five billion years ago on the planet Omniscia—one of a five-planet system in orbit around the star Draziw, two million light years away in the Andromeda galaxy. Merlin's birth coincided with the formation of the solar system in the Milky Way galaxy that contains planet Earth.

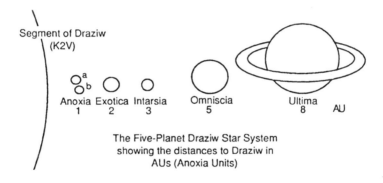

Segment of Draziw
(K2V)

a					
b					

Anoxia Exotica Intarsia   Omniscia   Ultima
  1   2   3   5   8   AU

The Five-Planet Draziw Star System
showing the distances to Draziw in
AUs (Anoxia Units)

Most residents of Omniscia are excited by science, and all residents have an endless thirst for knowledge. Merlin earned degrees in astrophysics, history, geophysics, chemistry, and philosophy at Omniscia's planetwide Universe-ity. Throughout Merlin's youth, Merlin was intrigued by the solar system that was born to the Milky Way when Merlin was born to Omniscia.

Merlin later took a keen interest in scientific thought as conceived by humans on the planet Earth.

Merlin noticed that humans were curious about the scientists' understanding of the universe but that many did not know where or how to find the answers to their questions. Sharing knowledge and wisdom is as fundamental as acquiring knowledge and wisdom. Merlin chose to visit Earth to spread science to those humans that share, along with Omniscians, that unquenchable cosmic thirst.

I

EARTH

*U*NLIKE MANY OTHER PLANETS in the universe, Earth is a dynamic place.

There are oceans of water that are tidally induced by the Moon to slosh upon the continental shelves. The continents slide determinedly upon a mantle that is the source of magma that erupts from volcanoes. There are plants and animals and microorganisms that live and reproduce in temperatures that range from the frozen Arctic to hundred-degree desert days. All this activity unfolds under nearly six quadrillion tons of a turbulent atmosphere that spawns storms, drought, electrical discharges, and erosion.

Curiously, this environment represents quite hospitable conditions when compared with other places such as Venus, Pluto, and, of course, the binary planet Anoxia in the Draziw system.

The tilted Earth rotates and wobbles and bobs its way through space in orbit around the Sun with the Moon tightly tethered nearby.

Dear Merlin,

What would happen if the Earth suddenly stopped rotating?

JONATHAN SWANN

SAN DIEGO, CALIFORNIA

*A*mong other things, we would all fall over and roll due east at about 800 miles per hour (the exact speed depends on your latitude on Earth). The Pacific Ocean would wash up onto North and South America and the Atlantic Ocean would wash up onto Europe and Africa. Many other unpleasant things would also happen.

Once it all settled down, the Earth day would equal the Earth year and there would be no tornadoes, hurricanes, cyclones, or typhoons.

Dear Merlin,

    If Earth were to explode tomorrow, what would happen to the orbits of the other planets?

        PATRICK KANE
        AUSTIN, TEXAS

$\mathcal{M}$erlin prefers to consider what would happen to *Earth*'s orbit if another planet were to explode tomorrow.

    The exact orbit of Earth is controlled by the Sun's mass and the mass of all remaining planets. An object's mass and its distance is all we need to know to completely determine the effects of its gravity on Earth.

    The Sun is a thousand times more massive than all planets combined, and is relatively close to Earth. This veritable cosmic arrangement will permit Earth's orbit—in the face of any other planet's armageddon—to remain within 99.999 percent of its current path around the Sun.

Dear Merlin,

What is the exact shape of the Earth? I'm told it is not a sphere.

ROBERT MCKINNEY
ALBANY, NEW YORK

*E*arth is slightly flattened at the poles and slightly wider below the equator than at the equator. This shape is unflatteringly referred to as an *egg-shaped oblate spheroid.*

Dear Merlin,

If lines of longitude on Earth help determine boundaries of time zones, and if all lines of longitude converge the farther north and south of the equator one travels, then what time is it at the poles where all lines of longitude meet?

DEAN JORDAN

MONTREAL, QUEBEC

I t is time to go back home.

There are no official time zones at the poles.

Dear Merlin,

As I understand it, Earth's core was very hot billions of years ago and it's been cooling ever since. Is it possible that the core will cool completely? If so, what would the consequences be?

JULIE JONES
BERKELEY SPRINGS, WEST VIRGINIA

Yes, Earth's core will one day cool "completely." When this happens, Earth's land masses will be geologically dead; no continental drift, no mountain building, no volcanoes, and worst of all—no hot springs.

Dear Merlin,

Suppose a hole were dug from one side of Earth, through the center, and out the other side. What would happen to a man if he jumped into the hole? When he got to the middle of the Earth would he keep falling or would he stop?

DEBBIE CANDLER

RED BUD, ILLINOIS

$\mathcal{H}$e would be vaporized by the 11,000° Fahrenheit temperature of the pressurized molten iron core.

Ignoring this complication, he would gain speed continuously from the moment he jumped into the hole until he reached the center of Earth where the force of gravity is zero. But he will be traveling so fast that he will overshoot the center and slow down continuously until he reached zero velocity at the exact moment he emerges on the other side.

Unless somebody grabs him, he will fall back down the hole and repeat his journey indefinitely.

A one-way trip through Earth would take about forty-five minutes.

Dear Merlin,

Does the Earth really wobble on its axis? If so, then why don't we notice it?

TOBY THURSTON
PATTERSON, NEW JERSEY

A full wobble (officially designated as the luni-solar precession) takes about 26,000 years. This is much too slow to be felt by anyone. If you wanted to observe the wobble then just come back in about 12,000 years. You will notice, if you look north, that the axis of Earth has "wobbled" away from Polaris. It will be pointing toward Vega—the polestar of the future.

Dear Merlin,

I heard that Earth's rotation is slowing down. Is this true?

ARNOLD BATES

SAVANNAH, GEORGIA

Yes.

The day gets about one second longer every 67,000 years. This is due to several effects—the most prominent of which is the oceanic tides that slosh back and forth on the continental shelves. The friction between the 1.5 quintillion tons of ocean and the land masses serves to dissipate some of the rotational energy of Earth.

The tides and sloshing and slowing of Earth's rotation will all end when the Earth day lengthens to equal the lunar month. In official jargon this is called "tidal lock."

Dear Merlin,

I would like to understand more about precession. Does the precession of Earth change as Earth's rotation changes? If so, does the precession period increase or decrease as the rotation of Earth slows? Also, do other solar system bodies precess?

NANCY HOGAN

LATHROP VILLAGE, MICHIGAN

$\mathcal{R}$otational precession is the "wobble" that results if a nonspherical object rotates at an angle under the influence of an external source of gravity. All planets fit this description. Earth, for example, isn't a perfect sphere and is tilted 23½ degrees on its axis and is tugged continuously by all other objects in the solar system, especially the Sun and Moon.

The equations of dynamics predict that the 26,000-year precession period will decrease by about one year for every three seconds that the Earth day is lengthened.

Dear Merlin,

What effect, all things remaining the same, does the great precessional cycle of Earth have on the weather?

FRANCIS M. BLOCK

ROBSTOWN, TEXAS

T he "great" 26,000-year precession cycle of Earth has no effect on the weather.

It does, however, affect what part of the year the seasonal nighttime sky will be seen. As Earth precesses, the constellations we associate with the various seasons (e.g., Orion in the winter, Cygnus in the summer) migrate through the calendar so that one half of a cycle from now, the June nighttime sky will contain the "winter stars" of December.

Dear Merlin,

I find the story of the precession of the equinoxes a fascinating one. Since this movement was detected in about 125 B.C. there must have been accurate records before then so: 1) When was the zodiac invented? 2) When was the vernal equinox at the first point of Aries? 3) What is the best date for the "Beginning of the Age of Aquarius"?

JACQUES L. SHERMAN, JR., M.D.
MIAMI BEACH, FLORIDA

The constellation names have origins in many cultures including the Chaldeans, Babylonians, and Egyptians of two to three thousand years ago. But it was circa A.D. 150 when the Greek philosopher Ptolemy first listed and delineated the twelve constellations that form the zodiac.

At that time, the annual path of the Sun against background stars was such that on the vernal equinox (the first day of spring), the Sun "entered" the constellation Aries. This is how the first day of spring became endowed with the name "first point of Aries."

Over eighteen hundred years later, a few things have changed. In 1930, the International Astronomical Union (IAU) restructured constellation boundaries—the annual path of the Sun now has fourteen constellations. And because of the ongoing precession of Earth on its axis the "first point of Aries" has shifted in the calendar and occurs one month earlier than the vernal equinox.

In about six hundred years the "first point of Aries" will reach the constellation Aquarius and enter the "age of Aquarius." While there have been songs written about this upcoming era there is no scientific reason to get excited about it.

Dear Merlin,

I just found out that Earth has a flattened orbit that brings it closest to the Sun in January and farthest from the Sun in July! How can this be? The seasons seem to indicate just the opposite of this.

PAM STARKEY

DALLAS, TEXAS

*E*arth is indeed closer to the Sun in January than in July. The seasons, however, have a very different cause.

Earth's axis is tilted 23½ degrees from the plane of its orbit around the Sun. When it is summer, the northern hemisphere is tilted toward the Sun. When it is winter, the northern hemisphere is tilted away from the Sun. You may have noticed that the midday sun in the summer is higher in the sky than the midday sun in the winter.

When the Sun is high in the sky the ground heating is much more efficient than when the Sun is low. The Sun heats the ground and, after a small time delay, the ground heats the air. This is why the hottest time of day is a few hours after 12 noon and the hottest time of year is one to two months after the summer solstice (June 21).

Of course, all seasons are reversed for dwellers in the southern hemisphere.

Dear Merlin,

If Earth did not have its 23½ degree axial tilt, that is, if its axial tilt were zero degrees, what effect would that have on our seasons? Would the northern and southern hemispheres have the same seasons concurrently?

ELVIS RAFFO

MORGANVILLE, NEW JERSEY

There would be no seasons, and every spot on Earth would get exactly twelve hours of daylight and twelve hours of night—every day would be an equinox.

You would also notice that bears wouldn't hibernate, trees wouldn't know when to lose their leaves, and fashion wouldn't know when to change.

Dear Merlin,

In the fall and winter we can get five or six UHF stations well, but shortly before the spring equinox the reception starts getting poor, and over a week or so we gradually get no UHF reception at all until the following fall. Does the tilt of Earth toward the Sun have anything to do with TV reception?

LEE LONG

NAPA, CALIFORNIA

If we assume that the workers at the five or six UHF stations don't take summer vacations every year then you are correct to suspect Earth's tilt.

In the summertime Earth's axis tilts its northern hemisphere toward the Sun thus increasing the Sun's interference when compared with the winter. If your UHF stations were weak to begin with then it is likely the Sun's interference was stronger than the UHF signal you wanted.

The "snow" you probably saw on your TV screen (and the "static" you hear between radio stations) is called *radio noise.* It comes from a variety of places like nearby electrical generators, appliances, power lines, but most importantly, the Sun. Earth receives more radio noise from the Sun than from any other object in the sky.

Dear Merlin,

I once read that the axis of Earth has flipped several times in the past. Can this possibly be true?

BRIAN OVERTON

OTTAWA, ONTARIO

No. Not if you're referring to Earth's axis of *rotation*. The magnetic poles, however, have flipped many times in Earth's history but there is still no consensus among geologists about how or why this happens.

Dear Merlin,

If most of the universe is hydrogen then how come Earth's atmosphere barely has any? Where do we get it from when we need it?

JESSICA WHITMAN
BALTIMORE, MARYLAND

A t the temperature of Earth's lower atmosphere, hydrogen atoms find themselves moving at speeds far in excess of Earth's 7 miles per second (25,000 mph) escape velocity. They escape to interplanetary space quite easily.

If you ever need some hydrogen, it can be found emerging from your nearest faucet. What comes out is an enormous number of molecules with two hydrogen atoms that are bound strongly to one oxygen atom. We call this chemical configuration "water." By methods of electrolysis scientists break the hydrogen–oxygen bond thus releasing hydrogen gas and oxygen gas.

Dear Merlin,

If the escape velocity on Earth is 25,000 miles per hour then how can the Earth have an atmosphere since molecules travel at speeds greater than 25,000 miles per hour?

DAVID MILLER

PITTSBURG, CALIFORNIA

I ndeed, the escape velocity on Earth is about 25,000 miles per hour. But the oxygen and nitrogen molecules at the surface of Earth (the fastest in the atmosphere) have an average speed of just over 1,000 miles per hour so you needn't worry about waking up in a vacuum one day.

Dear Merlin,

I have a very simple yet mystifying question for you. Why is it that when you leave the atmosphere it is cold, but coming back it is hot?

ADAM BENKOSKI

AUSTIN, TEXAS

Your temperature in space depends on what you do with the sunlight that hits you.

If you plan to take such a round-trip then you can control your temperature in space simply by the color of clothes you wear. But if you wear black, you will *absorb* all the Sun's rays that hit you and then your body temperature will rise to 270° Fahrenheit. Since this is somewhat above the boiling point of blood you might consider wearing different color clothes.

If you wear clothes with mirrors riveted all over them, you will *reflect* most of the Sun's rays that hit you, you will die, and your temperature will drop past 100° below zero.

The ideal way to leave Earth's atmosphere is to wear what astronauts wear—a temperature-controlled space suit.

Dear Merlin,

    Why is air thinner on mountaintops than at sea level?

      VIRGIL NICHOLSON

      LITTLE ROCK, ARKANSAS

*U*nlike solids and liquids, air is very compressible. Air pressure is determined by the weight of an entire column of Earth's atmosphere that is above a chosen area. When you go to a tall mountaintop (as observational astronomers do routinely) you have left thousands of feet of air column below you. The remaining air above you doesn't weigh as much as before so it compresses the air around you less.

Some effects of this phenomenon include the adjustment of cooking times for food because the exact boiling point of water (a major food additive) depends on the surrounding air pressure. If there are ever restaurants on the Moon they will surely have an unusual menu—but you can guarantee they will have no atmosphere!

II

MOON

MOON

HE MOON IS Earth's lone natural satellite. When viewed from anywhere on Earth, it is above the horizon just as often in the daytime as at night. Of course its singular splendor is enhanced at night when it does not share the sky with the Sun.

The Moon's moonthly journey in orbit around Earth permits us to view the lunar surface as it is illuminated by sunlight in a continuous range of angles. These are commonly called "phases" that progress from the invisible *new moon* (where the Moon is between the Earth and Sun and the far side is illuminated completely), to *waxing crescent*, to *first quarter* (commonly called half moon), to *waxing gibbous*, to *full moon*, to *waning gibbous*, to *last quarter*, to *waning crescent*, and back to *new moon*. The last quarter and waning crescent moons rise after midnight so these two phases tend to be appreciated only by all-night security guards, and "moonlighting" taxi cab drivers.

Two humans first landed on the Moon July 20, 1969. Had they traveled at a law-abiding 55 miles per hour, it would have taken about seven months, but rocket propulsion permitted the trip to occur in less than seventy-two hours. The astronauts found what was expected—a barren, waterless, airless, cratered surface. They got to hop around and plant a flag and collect rocks and leave behind their footprints in the dust of lunar time.

Dear Merlin,

I have noticed in various newspapers that the exact minute the Moon is full is always given. What I would like to know is exactly how long is the Moon full?

DORIS BRINLEE
DALLAS, TEXAS

The definition of a full moon is the instant the Moon is opposite the Sun in the sky as seen from the center of Earth. The time given in newspapers and publications is the nearest minute that this occurs. The Moon will appear quite full for the entire night.

To the untrained eye, however, the Moon can seem to be full for several days before and after the official date, until it ambles through to its next phase, "waning gibbous."

Dear Merlin,

I once read that it was impossible for there to be two full moons in February. But every four years February has 29 days. So wouldn't it indeed be possible—"once in a blue moon"—for there to be one full moon on February 1, and one on February 29?

CLAIRE P. GREENE

OCONOMOWOC, WISCONSIN

The average time between consecutive full moons is 29 days, 12 hours, 44 minutes, and 3 seconds. February can never provide more than 29 days, 0 hours, 0 minutes, and 0 seconds. It cannot have two full moons.

For February then, the expression "once in a blue moon" means never.

Dear Merlin,
   How bright is moonlight on a full moon night?
      KIP CONNELL
      REDWOOD CITY, CALIFORNIA

The Moon is almost as bright as the suburban streetlight that you may be standing under—yet it looks that bright after sunset for everybody on Earth. It creates shadows, "drowns" out the light of thousands of normally visible nighttime stars, inhibits the view of seasonal meteor showers, and aids in burglaries. But it also permits a harvest to continue past sunset, saves flashlight batteries when you're camping, and turns an ordinary evening into a romantic one.

For all of this brightness, the Moon reflects only about 7 percent of the sunlight that strikes its surface. The rest is absorbed.

Incidentally, "full earth" seen from the Moon is over fifty times brighter than full moon seen from Earth.

Dear Merlin,

Could you provide the names for all the full moons in the year?

JIM RICHARDS

LIVERMORE, CALIFORNIA

*F*ull moon names differ from country to country and from region to region. Below is a list of the more common names given to moons as described in North America.

The January moon is unimaginatively named the *Moon After Yule*. The February moon is the *Snow Moon* (if you happen to be wallowing in snow), or it is the *Wolf Moon* (if you notice wolves that howl at it), or it is the *Hunger Moon* (if you ran out of the food reserves from your autumn harvest.) As every tree knows, the March moon is the *Sap Moon*. In April it is the *Grass Moon*. And in May it is the *Planting Moon*. The full moon doesn't seem to know about the rhyme "April showers bring May flowers" because "in June we get the *Flower Moon*." The June full moon also stays low in the sky. It often takes on a honey-amber color from the atmospheric dust low on the horizon. Consequently, newlyweds prefer to call the full moon in June the *Honey Moon*. The *Thunder Moon* is appropriately found in July while August has the *Grain Moon*. The closest full moon to the autumn equinox is the *Harvest Moon*. This full moon rises just after sunset so one can harvest through the night. If the Harvest Moon is in October, then the September moon is the *Fruit Moon*. And if the Harvest Moon is in September then some animals will be dismayed to know that the October moon becomes the *Hunter's Moon.* The November moon is the *Frosty Moon,* especially if you live in the North. And the December moon is called *Moon Before Yule* (unless it falls after Christmas where it is then called *Long Night Moon.)* A second full moon in any month is always called the *Blue Moon.*

This fatiguing list of full moon names was obviously invented in an older, agricultural period of North American society. Now that times have changed, Merlin would like to suggest a more appropriate revised list of full moons.

January:	Super-Bowl Moon
February:	Dirty-Snow Moon
March:	Weather-Is-Getting-Better-Moon
April:	Tax-Return Moon (if before the 15th)
	Late-Fee Moon (if after the 15th)
May:	It's-Getting-Warmer Moon
June:	Elope Moon
July:	Muggers'-Night Moon
August:	Muggy-Night Moon
September:	Back-to-School Moon
October:	Fallen-Leaf Moon
November:	Bare-Trees Moon
December:	White-Christmas Moon (if you live in the North)
	I'm-Dreaming-of-a-White-Christmas Moon
	(if you live in the South)

Dear Merlin,

Does the full moon affect people's behavior?

AMY CANTOR
CHICAGO, ILLINOIS

In some municipalities more babies are born during full moons than any other phase.

The burglary rate in large urban areas (e.g., your home town) goes up during full moons.

But before we jump to cosmic conclusions we must look at these two statistics more closely. The human gestation period is very nearly equal to ten cycles of the lunar phases. The fact that more babies are born during a full moon simply means that more babies are conceived during a full moon. And nobody will argue the romantic effects of a moonlit evening.

The full moon is the only phase that is visible all night. It rises at sunset and sets at sunrise. It is also the brightest phase. Burglars know all about this and attempt to take advantage of these ideal lighting conditions throughout the night. During cloudy nights—when there is a full moon—the burglary rates are no different from any other time of the month.

P.S. Merlin has never seen anyone grow hair on their palms and fangs in their mouths during *any* phase of the Moon.

Dear Merlin,

How come during a thin crescent moon you can sometimes see the outline of the rest of the Moon?

MARY DUGGAN

CORPUS CHRISTI, TEXAS

This phenomenon was first described correctly by Merlin's good friend Leonardo da Vinci in the late fifteenth century.

During a thin crescent moon the Earth–Moon–Sun alignment permits a "moon being" on the darkened side of the Moon to see full Earth. It appears as a cloud-draped blue ball nearly fourteen times larger in the sky than the full moon viewed from the Earth.

The full Earth provides enough light on the darkened lunar surface to be visible back on Earth as a faint outline of the rest of the Moon. Astronomers call this phenomenon "earthshine"—although Merlin prefers "moonshine."

Dear Merlin,

I am intrigued when I look at the Moon when it is not full, and I see a faint outline of the darkened portion. Is this a part of the lunar hemisphere that we never see (the dark side of the Moon), or is it the same lunar hemisphere always facing us with the position of the lunar dawn simply moving across it?

JIM TODD
SAN DIEGO, CALIFORNIA

Contrary to popular musical literature and folklore, there is no "dark side" of the Moon. Indeed, the position of the lunar dawn migrates across the entire lunar surface to provide nearly fifteen consecutive days of sunlight to every part of the Moon. In scientific circles, this moving boundary between light and dark has the less-than-poetic name, "terminator."

The Moon does, however, display only one side—the "near side"—to the Earth at all times. In late 1959, the Soviet spacecraft *Luna 3* flew past the Moon. Only then did Earthlings obtain the first photographs of what the "back side" of the Moon looks like.

Dear Merlin,

When we were in China one of our group declared that the Moon (it was a crescent) was facing in the opposite direction from the way it would be at home. I thought that must be wrong, but I didn't know enough to argue.

I have just received a letter from a friend who has been in New Zealand. She says, "the Moon fascinated me. In New York it was in the first quarter . . . [when] we were in the southern hemisphere the next night it looked as if it were in the last quarter."

Can you straighten me out on all this?

    EDITH F. RIDINGTON
    WESTMINSTER, MARYLAND

The Moon should look no different in China than in the United States.

During a visit to the southern hemisphere, however, all celestial objects (planets, moons, constellations, etc.) that you previously determined to be "right-side-up" will appear "up-side-down." When you turn something up-side-down (like your point of view), the right side flips to the left and the left side flips to the right.

The true phase of the Moon doesn't care much about whether you were flipped when you looked at it.

Dear Merlin,

How big are the craters on the Moon?

CORY JOLLY

AUSTIN, TEXAS

The Moon's surface has wider craters, deeper valleys, and longer ridges than any corresponding feature on Earth's surface.

The lunar "Highlands" contains craters of all sizes up to 200 miles in diameter with walls that rise up to 10,000 feet above the surrounding terrain.

With safety in mind, the *Apollo* missions to the Moon naturally targeted the flattest possible areas for their landing sites.

Dear Merlin,

Why doesn't the Moon have an atmosphere while Earth, so nearby in space, has an appreciable one?

SCOTT McGRUDER

FORT COLLINS, COLORADO

The nitrogen and oxygen air molecules in Earth's lower atmosphere travel at about 1,600 feet per second between their random collisions. Earth's gravity is sufficiently strong to prevent molecules like nitrogen and oxygen from escaping into space despite their high speed.

The Moon's feeble gravity (one sixth that of Earth) succeeds in keeping only the slowest of gas molecules. As a consequence, the "atmospheric" pressure on the Moon is only one trillionth that of Earth.

Dear Merlin,

If the Sun's gravity is stronger than Earth's gravity then why does the Moon orbit the Earth?

ROY SPARKMAN

ALBANY, NEW YORK

$\mathcal{U}$pon closer inspection you will see that while Earth and the Moon are in mutual orbit, they *both* orbit the Sun together.

Dear Merlin,

Astronomers say that the Moon moves away from Earth a little bit per year. Why is this so?

JULIE JONES

BERKELEY SPRINGS, WEST VIRGINIA

The Moon's gravity exerts a "tidal" force on Earth that (among other things) slows Earth's rotation.

In response to this loss of rotational momentum by Earth the Moon increases its momentum of revolution by moving farther away from Earth at a rate of about an inch per year.

This is in accordance with a more general principle of physics called the "conservation of angular momentum."

Dear Merlin,

What was the maximum velocity of the Apollo spacecraft on their way to the Moon and how long was the actual time of transit between the Earth and Moon?

LEE DERR

FLORISSANT, COLORADO

The Apollo spacecrafts all averaged about one mile per second for their three-day journey from Earth to the Moon. The maximum speed for all of them occurred just after leaving Earth orbit at the beginning of their trip. This is where they fire their engines to reach Earth's escape velocity of seven miles per second. From that moment onward they "coast" toward the Moon (provided they are aimed properly) and are continually slowing down due to Earth's gravitational pull.

When they are within 27,000 miles of the Moon, the Moon's gravity becomes strong enough to increase their speed until they choose to enter lunar orbit.

Dear Merlin,

Could you tell me where to look on the Moon to find the exact spot that the first astronauts planted the American flag?

BETTY RAY DUNBAR
SOUTH GATE, CALIFORNIA

*A*t last count there are six American flags in the Moon left there by the *Apollo* astronauts. They are all in different places. The first flag, planted by Neil Armstrong in 1969, was placed near their landing site in Mare Tranquilitis—the Sea of Tranquility. It is a flat, wide-open section of the lunar surface.

While the Sea of Tranquility is plainly visible from the surface of Earth with the unaided eye, the actual flag is too small to be seen by even the largest ground-based telescopes.

Dear Merlin,

What is the best resolution of lunar detail that can be seen by the 400-inch telescope here on Earth? That is, can craters one mile in diameter be discerned? Half a mile? (I've read this before, but cannot find it again in my 150 astronomy books.)

WILLIAM R. DELLINGES, STARGAZER

NEWARK, CALIFORNIA

The 400-inch Keck telescope in Hawaii can resolve craters down to about a mile in diameter. But before you get impressed, you should know that a four-inch telescope would provide the same resolution. The resolving limit in this example is determined by Earth's turbulent atmosphere.

If you hold your breath while you lift the telescopes above Earth's atmosphere you will notice that the four-inch telescope will still resolve only one-mile craters but the 400-inch telescope will see detail down to fifty feet.

Dear Merlin,

What are the first words spoken from the Moon? I've heard conflicting stories about this.

JONATHAN MARSHALL
HOUSTON, TEXAS

$\mathcal{A}$ s many Texans know, the first word of the first comments spoken from the Moon is HOUSTON. Merlin was visiting the Moon when all this happened and overheard the following dialogue between *Apollo 11* astronaut Neil Armstrong and Mission Control.

ARMSTRONG: *Houston, Tranquility base here. The Eagle* [lunar module] *has landed.*

MISSION CONTROL: *Roger, Tranquility, we copy you on the ground. You've got a bunch of guys about to turn blue. We're breathing again. Thanks a lot.*

ARMSTRONG: *Thank you.*

MISSION CONTROL: *You're looking good here.*

ARMSTRONG: *A very smooth touchdown.*

III

PLANETS

P_{LANETS}

THE SUN'S NINE PLANETS (and Draziw's five) are all so different from each other that many astronomers devote their life's research to a single planet. When broadly categorized, the solar system has four small, rocky planets—Mercury, Venus, Earth, and Mars; four large gaseous planets—Jupiter, Saturn, Uranus, and Neptune; and one planet, Pluto, that is in a class by itself.

Unlike comets, all planets orbit the Sun in roughly the same plane and in the same direction (counterclockwise when viewed from the "top"). These two important facts hint to a common dynamical origin for all planets. Current theories suggest an enormous rotating gas cloud that collapsed and flattened as it rotated more and more quickly. The Sun, with 99.87 percent of the mass, formed in the center while the nine planets condensed around it in a common orbital plane. There also exists an outermost portion of the original cloud that did not partake in the collapse and flattening. It remains in the cold of distant space as the Sun's primary source for comets.

Dear Merlin,

Could you list all the planets in order from the Sun for me?

V. SCHWARTZ
TRENTON, NEW JERSEY

Merlin always remembers the planets in order of the distance from the Sun with the following mnemonic: "My Very Educated Mother Just Served Us Nine Pizzas." The first letters match the first letters of the planets in order: Mercury, Venus, Earth, Mars, Jupiter, Saturn, Uranus, Neptune, Pluto.

But from 1979 to 1999, Pluto, in its elongated orbit, was closer to the Sun than Neptune. Between those years, our mnemonic actually reads: My Very Educated Mother Just Served Us *Pizzas Nine.*

Dear Merlin,

I believe the planets are all named by the Greeks—or, at least, with Greek names. The name, Earth, doesn't sound Greek. How was our planet named?

SHIRLEY Z. HARTWELL

GUILFORD, CONNECTICUT

"Earth" comes from the Old English *eorthe* meaning "ground."

Note that the word for your planet, strictly speaking, should not be capitalized because it isn't named after anybody. For that matter, neither is the "Moon." All other planets of the solar system, and their moons, are named after miscellaneous characters in Roman and Greek mythology, and in Shakespearean plays.

Dear Merlin,

What are the most difficult obstacles to overcome for a manned exploration to the neighboring planets?

RON SIMPSON
CHRISTINA, PENNSYLVANIA

Mercury and Venus have surface temperatures in excess of 700° Fahrenheit. These inhospitable conditions are sufficient to melt lead and zinc. Venus poses an additional problem because it has nearly 100 times the atmospheric pressure on Earth. Space suits would need to withstand 1300 pounds per square inch.

The periodic raging dust storms on Mars can be avoided if you land at the right time and place on the Martian surface.

If you manage to traverse the asteroid belt with no major structural damage to your spacecraft then you will discover that the gaseous planets, Jupiter and Saturn, offer no "surface" to land on—no place to plant a flag. Also the excessive gravity on Jupiter would make a 160-pound astronaut weigh over 400 pounds.

Uranus, Neptune, and Pluto are each a bone-chilling 350° below zero Fahrenheit. Other dangers in these remote planets remain to be determined.

The most difficult obstacle of them all, however, is funding.

By far, then, the safest and most affordable planet for people to visit is Earth.

Dear Merlin,

Why does Mercury have so many craters while the Moon has comparatively few? Both Mercury and the Moon have no atmosphere.

DANNY NEGVESKY
MANASSAS, VIRGINIA

The side of the Moon that never faces Earth, "the far side," looks just as cratered as Mercury. This fact wasn't established until the mid-sixties when the Soviet Union sent a lunar orbiter to photograph the Moon's far side.

The entire lunar surface was once as heavily cratered as the far side. The subsurface of the near side of the Moon, however, was where regions of lunar lava were concentrated. The emerging lava filled these gigantic impact craters, and cooled to a relatively flat and smooth surface. After the formation of the solar system not enough asteroids and meteors remained to thoroughly "pepper" these surfaces with fresh craters.

Dear Merlin,

I've heard so often that Venus is Earth's "sister" planet and that they are very similar. How then do scientists account for Venus's surface temperature of 900° Fahrenheit? Is this so because Venus is closer to the Sun?

ROYCE VETTER

HARRISBURG, PENNSYLVANIA

The beautiful planet Venus has the unfortunate distinction of being the hottest planet in the solar system. Its surface temperature of 900° Fahrenheit is much hotter than a pizza oven. At 55° Fahrenheit, the average surface temperature on Earth is considerably cooler.

Venus is slightly closer to the Sun than Earth is. We would expect it to be only slightly hotter. The excess heat is the result of a runaway "greenhouse effect" caused by the large quantities of carbon dioxide in Venus's atmosphere. Visible light penetrates the thick cloud-cover of the planet and is absorbed by the rocky surface. This radiation is then reemitted in the form of infrared rays that get trapped by the carbon dioxide and heat up the atmosphere.

Earth's atmosphere is mostly nitrogen and oxygen with only small amounts of carbon dioxide. By comparison, Earth's greenhouse effect is meager.

Dear Merlin,

How can Venus's surface sustain the equivalent of ninety Earth atmospheres of air pressure? That's 1,300 pounds per square inch! That "air" must be dense or deep or both. It hardly seems that a planet can support that much pressure. What is going on?

BILL HAGEMEIER

DEL RIO, TEXAS

There is no problem for a rocky planet such as Venus (or Earth, its "twin") to support 1,300 pounds per square inch. The outer 200-yard layer of Venus's crust applies about the same pressure on the inner parts of the planet as the dense Venutian atmosphere that is hundreds of miles thick.

When you're made of rocks this kind of pressure doesn't bother you.

It is alarming to note that Earth's atmosphere would have even greater pressure than Venus if all the oceans were evaporated and all the carbon that is "locked" in life forms were released. Such a fate is possible through global nuclear holocaust.

Dear Merlin,
   Why is Mars red?
      HOPE REYNOLDS
      SAN ANTONIO, TEXAS

The surface of Mars contains iron oxide particles (more commonly called "rust") that are mixed with the other surface constituents.

It is this planet's distinct red color that inspired the Romans to name it after Mars, their god of war.

Dear Merlin,

Why is Mars so very cold when its atmosphere is mainly carbon dioxide, just like the atmosphere of Venus?

DAVID MILLER

PITTSBURG, CALIFORNIA

T he not-so-pleasant atmosphere of the "beautiful" planet Venus has fifteen thousand times the pressure of Mars's atmosphere. This bone-crushing environment forces every cubic inch of Venus's lower atmosphere to contain ten thousand times the quantity of carbon dioxide than the corresponding cubic inch of Mars's lower atmosphere.

The carbon dioxide on both planets traps heat, but you can see now why Venus is much more successful.

We also must not forget that Mars orbits, on average, twice as far from the Sun.

Dear Merlin,

In astronomy class we learned that because of the pressure within Jupiter's interior, astronomers theorize that hydrogen takes on the unfamiliar form of metallic hydrogen. I want to know if metallic hydrogen has ever been made (isolated) on Earth under laboratory conditions or is this metallic hydrogen some invention of the mind?

ERIC VANGERUD

VALLEY CITY, NORTH DAKOTA

$\mathcal{D}$eep within Jupiter the pressure is great enough to convert solid molecular hydrogen to solid metallic hydrogen. The metallic properties of the pressurized hydrogen are believed to be responsible for the intense magnetic field around Jupiter.

This rare form of hydrogen was first produced by a team of scientists in Livermore, California, in 1972. The results from their experiment indicated that the inner regions of Jupiter are indeed under enough pressure to produce metallic hydrogen.

Dear Merlin,

I recently read that the planet Saturn is light enough to float on water. Since the planet is so much bigger than the Earth how can this be true?

BONNIE VICKERS

FORT WAYNE, INDIANA

*I*ndeed, if you found a bathtub that was big enough to hold it then Saturn would float.

The jovian planets (Jupiter, Saturn, Uranus, and Neptune) are only about a third as dense as the terrestrial planets (Mercury, Venus, Earth, and Mars). It just happens that the density of Saturn, unlike the density of any other planet, falls below the density of water. When this condition exists for any object, cosmic or otherwise, it is guaranteed to float.

Dear Merlin,

I recently read that Saturn's rings disappear when viewed edge-on from Earth. Why is this so?

RALPH CARRERA
BRONX, NEW YORK

Saturn's rings are estimated to be only a few dozen miles thick. When viewed edge-on at Saturn's enormous distance it is no surprise that they seem to disappear. It is like detecting a dime at a distance of four hundred miles.

When Merlin's good friend Galileo Galilei first observed the "disappearance" of Saturn's rings in 1610, he made the mythological reference, "Can Saturn have swallowed his children?"

Dear Merlin,
I understand that occasionally Neptune is the farthest planet in the solar system. When will Pluto regain that distinction?
    DON LYLES
    OAKLEY, CALIFORNIA

*T*he questionable honor of being the most "way out" planet in the solar system will once again be regained by Pluto at 29 minutes and 19 seconds after 6 A.M. EST on February 10, 1999.

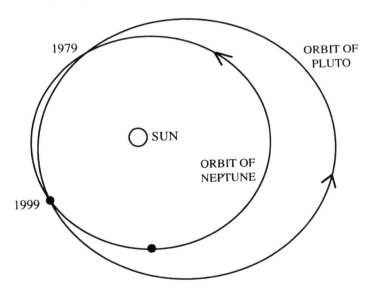

Pluto's average distance from the Sun is greater than that of Neptune. Pluto's flattened orbit, however, brings it closer to the Sun for twenty years out of its 248-year orbital period.

Dear Merlin,

What is Pluto: a planet, planetoid, or comet? How will it be determined if Pluto should be demoted to asteroid status?

ROY KRAUSE

SHAW AFB, SOUTH CAROLINA

*M*erlin has noticed over the years that many people would like to demote Pluto to an "–oid" status.

But Pluto is twice the size of Ceres, the largest known asteroid, and fifty times the size of the largest comets. When we consider that Pluto has a satellite of its very own it certainly gets Merlin's vote for full rank and privileges of "planet."

Dear Merlin,

How did astronomers know how long it took Uranus, Neptune, and Pluto to make one trip around the Sun after they were discovered?

DAVID HUTSELL

MODESTO, CALIFORNIA

$\mathcal{M}$erlin visited the German mathematician Johannes Kepler at his home back in 1619. It was soon after he formulated his third law of planetary motion.

Merlin asked, "Johannes, of what use is this law you have just formulated?"

Kepler responded humbly, "Merlin, with this law you may compute the orbit period of all five planets, plus the Earth, of course. All you need to know is a planet's average distance to the Sun."

"Then what?"

Kepler continued, "According to my third law of planetary motion, the orbital period is proportional to the square root of the average distance to the Sun raised to the third power."

"How did you determine this?"

Kepler recounted in a tone of frustration, "Merlin, I spent fifteen years working on this orbit problem. I kept presupposing that there be some mathematical connection between the *five* geometric solids and the *five* planets in the sky."

Merlin looked on with puzzled astonishment, "?!"

Kepler resumed, "I finally abandoned that idea because, to put it simply, it didn't work. The planetary data that I used were excellent. They were willed to me by Tycho Brahe, the great observational astronomer of Denmark. At Uraniborg, Tycho's observatory, he meticulously recorded the motions of planets across the celestial sphere. I finally let the data *tell me* how the solar system works rather than vice versa. My third law is the natural result."

Merlin commented, "It's an impressive law."

Kepler suggested, "It may even apply to planets that are yet to be discovered."

Little did Kepler know that his law would apply to all planets, all asteroids, all comets, and all debris, in orbit around the Sun. When Uranus, Neptune, and Pluto were discovered it became a simple matter to compute their orbital periods once you determined their average distance to the Sun—a much simpler method than waiting with your stopwatch for a couple of centuries.

Dear Merlin,

Why do the planets Venus, Uranus, and Pluto rotate in a retrograde manner? Could there possibly be a connection between these distant planets?

ROY KRAUSE
SHAW AFB, SOUTH CAROLINA

It is very likely that these three planets were each slammed by a large asteroid during the early days of the solar system. Such an event, if energetic enough, could easily upend a planet's axis and leave it with a "retrograde" rotation.

Dear Merlin,

When our Sun becomes a red giant, at the end of its lifetime, and it swallows all the planets up to Mars, what will happen to all the planets from Jupiter to Pluto?

CHRISTOPHER KUHLOW
NEWBURGH, NEW YORK

The gaseous titans of the solar system, Jupiter, Saturn, Uranus, and Neptune, are likely to evaporate away their tremendous atmospheres into interplanetary space. They will lay bare a tiny solid core of heavy elements about the size of Earth.

Some of the latest theories about the composition of Pluto describe it as a ball of frozen methane (the gas in household stoves), ammonia, and water. If this is true, then Pluto may evaporate away completely!

Dear Merlin,

What is the largest moon in the solar system?

JIMMY BAILEY, AND THE FLETCHER FAMILY

HOT SPRINGS, ARKANSAS

*I*t's a photo-finish between Jupiter's *Ganymede* at 3,270 miles in diameter, and Saturn's *Titan* at 3,200 miles in diameter.

Earth's moon (the only moon in the solar system with the name Moon) comes in at sixth place with a diameter of 2,160 miles.

Dear Merlin,

What is the latest count for the number of moons of each planet in our solar system?

STEWART R. BAKER

BLACK MOUNTAIN, NORTH CAROLINA

There are sixty known moons among the nine known planets.

Mercury has no moon, and
Venus does not, it is true.
But Earth, of course, has one, while
Planet Mars, take note, has two.

Mighty Jupiter, by jove,
Displays sixteen moons—what gall!
But Saturn sets the record
With seventeen—large and small.

Uranus has quite a few
With its fifteen moons in thrall.
Neptune's eight, and Pluto's one
Tallies sixty moons in all.

Dear Merlin,

On August 15, 1987, there was a well-publicized event referred to as the "Harmonic Convergence." Apparently, it is a rare astronomical event that occurs every 26,000 years in which humankind was meant to pray for peace. I have heard that the Aztecs predicted this event and that it should be of great significance. Is this a true astronomical rare event or just a hoax?

DANIEL G. LEVITT
TUCSON, ARIZONA

*A*n event can be rare without being interesting. For example, at *any* instant, the relative orientation of all planets will not repeat for a trillion years. By this measure, every day is "rare."

The day you select to celebrate, pray, worry, or chant, can be given any significance you choose, but on August 15, 1987, there was nothing astronomically unusual or interesting.

IV

Asteroids,
Comets,
and
Meteors

ASTEROIDS, COMETS, AND METEORS

*A*STEROIDS, COMETS, AND METEORS all have one feature in common—they are the "leftovers" from the original gas cloud that formed the Sun and planets. Asteroids are craggy chunks of rock. Comets are balls of dirt, ice, and frozen gases. And meteors are, quite simply, whatever falls through and burns up in Earth's atmosphere from outer space.

Asteroids are everywhere in the solar system but are found mostly in a crowded belt in orbit around the Sun between Mars and Jupiter. In keeping with the astronomer's tradition of literary simplicity, this belt of asteroids is called the "Asteroid Belt."

Most comets owe their origin to a distant "cloud" of frozen debris far beyond Pluto. The existence of this cloud was first suggested by the Danish astronomer Jan Oort. It has been aptly named the "Oort Cloud." Occasionally, for reasons not well understood, a comet will journey inward toward the Sun in an elongated orbit that brings it in view for earthlings to see.

Asteroids and comets are named by their discoverers. If meteors were visible for longer than the one or two seconds it takes them to burn up in the atmosphere then maybe meteors would have names as well.

Dear Merlin,

How many asteroids are there? What are their names? Are there more asteroids than planets?

ARNOLD GREEN
TUCSON, ARIZONA

There are hundreds of thousands of asteroids in and around the asteroid belt. But newly discovered asteroids are catalogued only if their orbits can be computed accurately. The last Merlin checked, the number of named asteroids was rising through five thousand, while there remained only nine named planets. Unlike the planets, however, every catalogued asteroid is also numbered.

Many asteroids are named for mythological people—

2101 Adonis,

2063 Bacchus,

55 Pandora,

1404 Ajax,

while others are named for mythological places—

1260 Walhalla,

1198 Atlantis,

2952 Lilliputia,

1282 Utopia.

Some are named for real cities—

3317 Paris,

2830 Greenwich,

2171 Kiev,
2224 Tucson,

while others are named for real countries.
469 Argentina,
1432 Ethiopia,
2575 Bulgaria,
2169 Taiwan.

Many are named after famous people that include the musicians
1814 Bach,
1815 Beethoven,
1034 Mozartia,
2266 Tchaikovsky,

and Merlin's good friends
662 Newtonia,
2001 Einstein,
697 Galileo,
1288 Santa.

Many others are named for some not-so-famous people—
1744 Harriet,
2335 James,
779 Nina,
1716 Peter.

But Merlin's favorite asteroid of them all is, of course,
2598 Merlin.

Dear Merlin,

I saw Halley's comet in 1910, a wonderful sight. What makes the comet return to its swing around the Sun after a journey of billions of miles into space? What has happened since 1910? Is there no other gravity in outer space to capture it?

A. D. McDEVITT
YOUNTVILLE, CALIFORNIA

$\mathcal{M}$uch has happened on Earth since 1910.

- Einstein develops the General Theory of Relativity (1915).
- Edwin Hubble confirms the existence of other galaxies and discovers the expanding universe (1920s).
- Clyde Tombaugh discovers Pluto (1930).
- Chuck Yeager breaks the sound "barrier" (1947).
- *Sputnik* is launched (1957).
- Neil Armstrong walks on the Moon (1969).
- *Venera* spacecraft soft-lands on Venus (1972).
- *Viking* spacecraft soft-lands on Mars (1976).
- *Pioneer* spacecraft "leaves" the solar system (1983).
- Comet Halley finds its way back to the Sun (1986).

We would have lost comet Halley if it wandered too close to the nearest star. But the nearest star is seven thousand times farther than the outermost part of comet

Halley's prescribed orbit. The Sun's gravity has a tight grip on all its planetary and cometary progeny.

When you see comet Halley for the third time in 2061, remember that between now and then the comet will have followed its flattened elliptical orbit, which always brings it back to the Sun.

Dear Merlin,

I read recently that in 2061 comet Halley would be five times brighter. Why?

 EDWARD SIDORSKI
SUSQUEHANNA, PENNSYLVANIA

The brightness of a comet in the sky depends on many things. But most important, it depends on the distance between the comet and Earth.

The dimmest arrangement is to have Earth on the "wrong side" of the Sun when the comet reaches perihelion. This happened with comet Halley's latest visit. In 2061 Earth will be a little closer to comet Halley at perihelion than in 1986. The comet will then appear brighter.

The brightest arrangement is for the comet to be at perihelion with Earth on the same side of the Sun. This happened in 1910 when Earth, incidentally, passed through comet Halley's tail.

Dear Merlin,

    With all the information available about Halley's comet, I've noticed no mention of other comets that also reappear. Could you come up with a top five comet list after Halley's?

    C. BROWN

    SUNNYVALE, CALIFORNIA

*T*here are two categories of comets. One type, called *periodic comets*, are those that return around the Sun in a "reasonable" length of time.

    Here is a list of some periodic comets.

Name	Designation	Period (years)
Encke	1786 I	3.3
Tempel(2)	1873 II	5.26
Faye	1843 III	7.41
Kearns-Knee	1963 VIII	9.01
Tempel-Tuttle	1866 I	32.91
Halley	1682 I	76.09

    Periodic comets don't tend to be very bright. In fact all but Halley in the above list are seldom visible to the unaided eye.

    The other type of comets, called *new comets*, are those that will either never return, or will return around the Sun in an unreasonable length of time—typically, many thousands of years. This is Merlin's favorite category because at least one or two new comets are discovered every decade that are visible to the unaided eye.

Some notable recent ones include those in the following list.

Name	Designation
Arend-Roland	1957 III
Humanson	1962 VIII
Ikeyi-Seki	1965 VIII
Bennett	1970 II
Kohoutek	1973 VII
Hale-Bopp	1995 O1
Hyakutake	1996 B2

If the statistics of comet discovery remain unchanged then a person living an average lifetime will see (without a telescope) comet Halley once, and from ten to twenty new comets.

Dear Merlin,

    If a periodic comet uses up some of its frozen center every time its tail is formed, what keeps it from getting smaller?

    L. DAY

    JOSHUA TREE, CALIFORNIA

*N*othing.

    Each time a comet rounds the Sun it loses mass in the form of particles and evaporated gases. It is the evaporated gases that are responsible for the comet's visible coma and tail.

    Some estimates for comet-mass loss have been as high as 1,000 tons per second for comet Kohoutek 1973 VII. With a total mass of hundreds of billions of tons Merlin expects many happy returns for comet Kohoutek. (You may miss the next visit because its orbital period is about five million years.)

    Short period comets, however, are feeble tail producers. They've been around the Sun so many times that rocks and dirt are all that remain.

    It is likely that some comets have broken up completely. The Geminid meteor shower cannot be attributed to any known comet and is believed to be the entrails of a comet that rounded the Sun once too many times.

Dear Merlin,

What causes meteor showers?

BARBARA J. IRVIN

AUSTIN, TEXAS

E arth, in its annual journey around the Sun, plows through over a thousand tons of interplanetary debris per day. As the debris enters the atmosphere and is attracted by Earth's gravity, it heats from friction and shock waves, glows, then disintegrates. When the particles that compose the debris glow bright enough to be seen with the unaided eye we call them *meteors*.

The side of Earth that happens to face the direction of orbit always bears the load of this daily encounter. (The hours between midnight and noon hold this distinction.) If it weren't for the protective blanket of the Earth's atmosphere we would all, no doubt, have meteor-collision insurance.

Occasionally, Earth will enter a part of its orbit where there is more debris than other parts. Excess debris is often associated with the residue of comets that have crossed Earth's orbit. These encounters are called *meteor showers* and are named for the constellation in the sky from which they appear to emanate.

To the right is a list of some prevalent annual meteor showers.

Name	Dates of Peak Shower	
Quadrantids	January	3/4
Lyrids	April	21/22
Eta Aquarids	May	4/5
Delta Aquarids	July	27/28
Perseids	August	11/12
Orionids	October	20/21
Taurids	November	8/9
Leonids	November	16/17
Geminids	December	13/14

Dear Merlin,

At dusk one evening, earlier this year, two friends and I observed a large fiery object in the sky. It soon broke into smaller pieces each of which left a long glowing streak across the sky. It was visible for about twenty seconds before it gradually faded out above the horizon. Did we see a meteor?

BETTY LERCH

MIDDLETOWN, OHIO

*M*erlin is glad you had other witnesses. Often mental anguish occurs when you are the only one to see a spectacular, rare event and nobody believes you when you tell them what you saw.

There is a type of meteor called a *bolide*. It is characterized by its fiery appearance and an explosion at the end of its streaking path. If the meteor starts out large enough it is reasonable that when it explodes it will break into smaller, glowing pieces. When meteors graze the atmosphere they are expected to last considerably longer (up to one minute) than their vertically falling cousins, which might be aglow for less than a second.

A more exciting prospect, however, would be that you witnessed the disintegration trail of a dead satellite reentering Earth's lower atmosphere. There are hundreds of artificial satellites that orbit Earth. In the growing history of satellite technology many have reentered.

Based on the description, Merlin's best guess is that you saw a bolide.

V

SUN

*T*HE SUN IS A RATHER ORDINARY STAR. It is not the biggest or the smallest or the hottest or the coolest of stars in the galaxy. In spite of this, the Sun is over a million times larger than Earth and is much too hot to visit. It is the direct and indirect source of all the energy content of life on Earth and all fossil fuels. In official astronomers' jargon the Sun is: Spectral Type G2, Luminosity Class V. (The cooler Draziw is K2V.) This coding indicates the color, temperature, and size of a star.

The Sun is a turbulent place. Close examination of its visible surface shows restless and unending change. High-energy charged particles are released constantly. The equator completes a revolution faster than the polar regions thus twisting and stretching the surface magnetic field lines. Dark blemishes called "sunspots" appear and disappear as they move across the Sun's surface. Gases in constant boil occasionally spew forth towering plumes of plasma called "prominences."

Surprisingly, all of this activity is quite tranquil when compared to what the Sun will do in five billion years.

Dear Merlin,
   How hot is the Sun?
      ELIZABETH YOUNG
      BALTIMORE, MARYLAND

On the Fahrenheit temperature scale, the Sun ranges from a rather uncomfortable 18,000,000° in its core to a tepid 10,000° at its visible edge.

The tenuous outer region of the Sun, which is commonly referred to as the "corona," has an estimated temperature in excess of 4,000,000°.

Dear Merlin,

I am told that the Sun's corona is millions of degrees while the visible surface is just 10,000° Fahrenheit. How can this be? One would think that the farther one is from a source of heat the cooler the temperature becomes.

ANDREW J. SMATKO, M.D.
SANTA MONICA, CALIFORNIA

Detailed studies of the Sun's spectrum have revealed rapid oscillations and pulsations in the visible edge of the Sun that send shock waves through the corona. Those shock waves serve as an efficient heating mechanism that maintains the rarefied corona at 4,000,000° Fahrenheit.

Dear Merlin,

How does the Sun make its energy?

NORMAN HESS

SAN DIEGO, CALIFORNIA

The Sun converts over five million tons of matter into energy every second in accordance with Einstein's famous equation, $E = m\ c^2$. (Energy equals mass times the speed of light squared.)

This veritable feat is accomplished in the 18,000,000° Fahrenheit core of the Sun via the "proton–proton" cycle in which the nuclei of four hydrogen atoms are converted to the slightly less massive nucleus of one helium atom.

Dear Merlin,

What is the "solar wind" and how fast does it travel through space?

C. B. SHEPHERD
TUJUNGA, CALIFORNIA

The Sun loses about one million tons of mass per second from its outer corona in the form of high speed electrons, hydrogen nuclei, and helium nuclei. These charged particles reach Earth at a speed of about 250 miles per second, and are collectively called the "solar wind."

It is the pressure of this solar wind that forces the tail of comets to point away from the Sun regardless of the direction the comet is moving.

Dear Merlin,

What are the sunspots doing these days? Ever since I read an article about the effects of sunspots on stock market prices I keep looking for data but seldom find it. Where are we in the current cycle?

A. E. BROWN
MIAMI, FLORIDA

People always try to associate earthly human events with cosmic happenings. Maybe it is because people don't want to be accountable so they pass the blame to the universe.

In the year 1997, the Sun was at "sunspot minimum." For the five or six years that follow, the Sun will approach "sunspot maximum" thus completing half of its eleven-year cycle.

Dear Merlin,

A while back I read a very interesting statement about our solar system. The article stated that over 90 percent of the mass of the solar system is present in the Sun. Yet over 90 percent of the angular momentum of the system is found in the planets. This seems to be an astounding statement. I would also like to know how a person can determine the angular momentum of the solar system. I would also like to know your explanation for this seemingly impossible statement.

> Jim Haglund
> Wheaton, Illinois

*A*ngular momentum can be represented as the mass of an object, multiplied by its orbital distance, multiplied by its orbital velocity:

$$angular\ momentum = mass \times orbital\ distance \times orbital\ velocity$$

The above equation works for circular orbits and becomes slightly modified for more flattened elliptical orbits. Each term on the right of the equal sign has a role in how big the angular momentum will be.

You can now see that even if your mass were small compared with the Sun, you can have a large angular momentum if your orbital distance is large enough. If we use the equation for each of the nine known planets and add the results, we compute easily the large part of the solar system's momentum that is blamed on the planets.

Dear Merlin,

Each year I notice that the Sun rises and sets at the same hour (A.M. and P.M.) several days after the autumnal equinox in September. Why were not day and night equal on the equinox? Why the delay?

ROBERT G. WEARNER

MOBERLY, MISSOURI

The equinox (literally—"equal night") is supposed to mean that the entire Earth gets twelve hours of light and twelve hours of darkness. Two important facts prohibit you from experiencing this phenomenon on the equinox.

1. The time of sunrise is the first appearance of the Sun above the horizon. The time of sunset should then be the moment the leading edge of the Sun dips below the horizon—but it is not. It is defined when the last bit of the Sun dips below the horizon.

2. Earth's atmosphere bends the sunlight (the way a half-submerged pencil appears to bend in a glass of water) in such a way that true sunrise and sunset occurred about five minutes before you saw them.

These two effects serve to "add" sunlight to the twelve hours of sun you are supposed to get on the equinox. "Visual" equinox is then expected to occur a few days after the official autumnal equinox and a few days before the official vernal equinox.

Dear Merlin,

What is solar retrograde motion? I heard about it recently but I'm not sure I understand it.

STEVE YOUNG

NEW YORK CITY

Every object in the solar system orbits the center of mass of the solar system. (If the solar system—Sun included—were embedded in a cosmic platter, the center of mass would be the point under which you would have to place your finger to balance it all.)

The center of mass is relatively close to the center of the Sun. Most of the time it is found deep within the Sun's surface. Normally, the planets and the Sun trace systematic loops around this center of mass. But the planets all have different masses, different distances, and different orbital speeds. About every 180 years we find that the arrangement of planets is such that one of the loops the Sun makes does not enclose the center of mass.

As seen from the solar system's center of mass, for about twenty years the Sun would appear to move in the opposite direction—"retrograde."

There are some theories that suggest this retrograde period will excite solar activity thus increasing sunspot counts above the expected numbers. This remains to be demonstrated.

Dear Merlin,

Is it just a coincidence that the size of the Moon is the same as the size of the Sun in the sky as viewed from the surface of Earth?

BOB DUPREE

PATTERSON, NEW YORK

$\mathbf{Y}$es.

Dear Merlin,

> *Why are solar eclipses considered so dangerous to look at?*
> NATHANIEL LLOYD, JR.
> CLARION, PENNSYLVANIA

The sun is always dangerous to look at with the unprotected eye.

During an eclipse, people might wish to run outside to look at the event without remembering to protect their eyes. Hence, the persistent warnings about the dangers of eclipses.

If, for some reason, people wanted to run outside every day to stare wide-eyed at the Sun then daily warnings would be just as necessary.

Dear Merlin,

If the Sun were to disappear all of a sudden, how much time would we have to think about it?

C. W. BARNES, JR.
DALLAS, TEXAS

If, as you read this, the Sun were to disappear, Earth (and you) would not know about it for five hundred seconds. Light and gravity, traveling at 186,282 miles per second, take this long to reach Earth's orbital radius.

At that moment, the daytime sky will go dark, the surface temperature will cool rapidly, and Earth will "fly" off at a tangent—lost in interstellar space.

In the five hundred seconds you had to think about your fate, you would *not* have known you should have been thinking about your fate . . . and then it's too late!

Dear Merlin,

I am told that when the Sun dies it will become a red giant. When this happens, what will happen to Earth?

G. FRAZIER
PISCATAWAY, NEW JERSEY

$\mathcal{A}$pproximately five billion years from now, when the Sun exhausts the hydrogen fuel in its core, we expect it to enter the red giant phase. The outer layers of the Sun will expand a thousand times their present diameter— engulfing the inner planets: Mercury, Venus, Earth, and Mars. The oceans of Earth will evaporate into space as the surface temperature rises to 5,000° Fahrenheit. Merlin estimates that these conditions will be wholly unsuitable for life.

There is no need to get terribly upset about all this because you won't live that long.

Dear Merlin,

   What would happen to Earth if the Sun became a supernova?
   TAMARA A. SMITH
   WASHINGTON, D.C.

It is fortunate for earthlings that the Sun will not ever become a supernova. Its mass falls far short of what is required to drive nuclear fusion reactions in its core from hydrogen and helium through to iron—the catalyst of many supernovae.

But if you still would like to imagine such a thing, the Sun would increase in luminosity a billionfold in an explosion of cosmic proportions. The inner regions of the Sun (more appropriately called the "guts" in this scenario) would be scattered throughout interplanetary space enriching it with the myriad heavy elements generated in the core.

Amid all this activity, Earth's surface would simply be vaporized and the ball that survived would be flung into interstellar space.

Dear Merlin,

    Planets orbit the Sun. As I understand it, the Sun has its own orbit around the galaxy that is perpendicular to the planets. How can this be?

      FRED L. RADIR
      ORLANDO, FLORIDA

T he correct picture is to consider that the Sun and the planets are in orbit around the center of the Milky Way galaxy at exactly the same rate. During this shared journey, the planets also happen to be in orbit around the Sun. Indeed, they are executing two orbits simultaneously.

    Since the plane of the solar system is nearly perpendicular to the path of the solar system around the galactic center, we can envision the planets to move along the coils of a cosmic helix (the shape of a spring) with the Sun moving along its center line.

VI

STARGAZERS

STARGAZERS

Look up dear friend,
What do you see?
I see autos and streetlights
And smog without end.

Higher dear friend,
Now do you see?
I see buildings and airplanes
And clouds now and then.

Up and out and
Beyond all of this
Is sky and space
You won't want to miss.

Now I see, I see,
I behold it all.
Was I blind all these years?
Did I miss my call?

I see the Sun as it sets
While twilight's curtain descends.

I see planets and clusters
And galaxies without end.

Meteors! And moonbeams!
More stars than the streetlights.
It all twinkles and gleams.
What a wonderful sight.

I'm now a StarGazer,
It's the source of my bliss.
'Cause I looked up and out
And beyond all of this.

Dear Merlin,

> *Why can't we see stars in the daytime?*
> MILT LARSON
> ATLANTIC CITY, NEW JERSEY

The Sun is very much a star and is plainly visible in the daytime.

The rest of the stars aren't bright enough. Much of the blue light that is present in the Sun's rays gets filtered out and scattered by Earth's atmosphere. This scattered blue light gives the daytime sky its light-blue glow, which prevents the detection of other stars.

On the Moon, where there is no appreciable atmosphere to scatter sunlight, the daytime sky is filled with stars.

Dear Merlin,
Why do stars twinkle?
F. BOWEN
WILMINGTON, DELAWARE

$\mathcal{M}$ost stars in the night are so far away that they should appear only as a tiny point of light that has traveled many trillions of miles to reach us.

But by the time this tiny point of light passes through the two hundred miles of ionosphere, ozonosphere, thermosphere, mesosphere, stratosphere, and troposphere the starlight is jiggled and wiggled and smudged and smeared by the different light-bending properties within and among these various layers of Earth's atmosphere.

When the light reaches your eyeball you are then likely to say, "Twinkle, twinkle, little star . . ."

Dear Merlin,

If Earth's atmosphere causes starlight to appear to twinkle, why does it not have the same effect on planets?

BRADLEY MORRISON

LEWIS, COLORADO

When starlight twinkles it is Earth's atmosphere that causes the "pinpoint" star image to jiggle and wiggle. This jiggle and wiggle forces a tiny change in the exact star position that is greater than the width of the star's image.

Planet images enter Earth's atmosphere as larger disks. The same tiny change in position is not as readily noticed since its extent is less than the size of the disk image. Twinkling for planets is therefore reduced significantly.

Dear Merlin,

How many stars are visible to the naked eye?

B. SCHULTZ
ROANOKE, VIRGINIA

There are anywhere between five and six thousand individual stars visible to the unaided eye in the entire sky. The exact number depends on how good your night vision is. Only half of these are ever visible at any one moment because standing on Earth prevents you from seeing the other half of the sky below you.

The number of visible stars goes up rapidly (from thousands to millions) with even the most feeble of pocket binoculars. Indeed, with the largest of today's telescopes the sky is the limit.

Dear Merlin,

If the Sun could be "turned off" on any given day, what would the daytime sky look like in terms of the identity and patterns of the stars and planets we could then see as compared to the night sky?

E. RAGER

DALLAS, TEXAS

*I*t is not very easy to turn off the Sun.

Merlin recommends that you: 1) choose a time and day of the year (like noon today) to go outside and look up at the clear blue sky; 2) go home for exactly six months; 3) go outside at midnight to look up at the star-filled sky.

That sky is the exact sky you couldn't see (except, perhaps, for a planet or two) when all you saw was "blue sky" six months before. As Earth orbits the Sun, the hidden daytime sky changes continuously as the Sun is superimposed on the different stellar backgrounds.

It takes exactly six months for the day sky to shift completely to the night sky and, of course, for the night sky to shift to the day sky.

Dear Merlin,

How come so many of the constellations in the sky don't look like the objects and animals they are supposed to represent?

B. MARSHALL

MINNEAPOLIS, MINNESOTA

The Chaldeans, Babylonians, Greeks, Egyptians, and Romans of two to three thousand years ago all share the blame for the celestial menagerie that remains today. Merlin is convinced they all had vivid imaginations. Especially this one fellow Merlin met in Corinth back in 600 B.C. He was the first person to label the constellation of Pegasus. He said he really did see an upside down white flying horse with wings in the sky.

Dear Merlin,

What constellations does the Sun pass in front of during the year?

J. R. MUSGROVE
OMAHA, NEBRASKA

C ontrary to popular belief, the Sun's annual path across the background stars intersects fourteen different constellations, not twelve. In order, beginning with the first day of spring, they are: Pisces, Cetus, Pisces, Aries, Taurus, Gemini, Cancer, Leo, Virgo, Libra, Scorpius, Ophiuchus, Sagittarius, Capricornus, and Aquarius.

Notice the Sun leaves Pisces for a while to pass through Cetus before it returns to Pisces again.

Dear Merlin,

How many constellations are there? It seems like there could be quite a few.

GERALD HARDY
CHEVY CHASE, MARYLAND

The International Astronomical Union officially recognizes eighty-eight constellations that cover the entire sky. Most of the names were first established over two thousand years ago.

Many are dim, like Corvus (the Crow) or Pavo (the Peacock). Some are boring, like the two-star constellation Telescopium (the Telescope), or the three-star constellation Triangulum (the Triangle). Others require extraordinary imagination to identify, like Apus (the Bird of Paradise) or Horologium (the Pendulum Clock).

The largest constellation is the long and winding Hydra (the Water Snake)—it covers over 1300 square degrees in the sky. And the smallest is Crux (the Southern Cross)—it covers a mere 68 square degrees.

Merlin would like to one day see all constellations updated to modern references with names such as: the *Buick*, the *Astronaut*, the *French Poodle*, the *VCR*, and the *Teddy Bear*.

Dear Merlin,

I recently bought a video recorder and would like to know how to use it to take pictures of constellations.

    C. ROBERT HICKMAN

    ATLANTA, VIRGINIA

Video cameras and motion picture cameras were invented to record images that were in motion. The individual stars of each constellation are certainly in motion, but due to their vast distances it would take you and your camera a few thousand years to notice it.

If you must videotape stars that don't move then be sure to use the widest aperture possible on the camera. You will successfully record some of the brightest stars in the most prominent constellations such as Orion, Cassiopeia, Sagittarius, Scorpius, and the Big Dipper.

Unlike regular cameras, video and motion picture cameras cannot increase their "exposure" time, which is critical for the accumulation of light from dim objects. Merlin therefore has doubts about the camera's ability to record the lesser constellations as well as deep sky objects like nebulae, star clusters, and galaxies, without the benefit of a telescope.

Dear Merlin,

   How did the "ancients" know that the Sun was in Capricorn at the winter solstice several thousand years ago?
   GEORGE RECORD
   NDANDA LINDI, TANZANIA

$\mathcal{M}$erlin is pleased to hear from readers in the southern hemisphere.

In modern times, the Sun enters the constellation Sagittarius about December 18. It remains there for the winter solstice, December 21, and doesn't enter the constellation Capricorn until about January 20.

Two thousand years ago, all constellations of the zodiac were shifted about one month from their present calendar positions due to Earth's precession on its axis. This would have put the Sun in Capricorn during the winter solstice and in Sagittarius during the preceding month.

Of course, the ancients couldn't see the stars of the constellation Capricorn when the Sun was passing through it. But a careful study of the stars in the east just before sunrise, and the stars in the west just after sunset on the winter solstice two thousand years ago would show the two constellations (Sagittarius and Aquarius) that flank Capricorn in the zodiac. By this method it was easy to infer for any day of the year what constellation the Sun must be in.

Dear Merlin,

In a map of the stars and the solar system that I saw recently, it looked like the Sun was in front of Gemini in July, in front of Cancer in August, etc. But the horoscope sign for late July is Cancer, and for August it is Leo. This differs from the way zodiac signs are matched to birthdays. Why?

BERNARD MAYOFF

RICHARDSON, TEXAS

The horoscopes that so many people are compelled to read were properly aligned with the position of the Sun over two thousand years ago when the Babylonians and the Chaldeans first divided the zodiac into twelve parts. A bit later, Ptolemy (circa A.D. 150) assigned the names of mythological figures and animals to the constellations that we know today.

Since then, however, the Earth has precessed on its axis by one twelfth of a full wobble. This shifts backward, by one month, the correspondence of the zodiac with the position of the Sun against background constellations.

There is also a prominent zodiac sign that doesn't even appear in the horoscopes. The Sun, after leaving Scorpius, passes in front of the constellation Ophiuchus. In fact, it spends more time in Ophiuchus than in Scorpius. A sobering conclusion for astrologers is that if you thought you were a Scorpio you are probably an Ophiuchan, and all Scorpios and Ophiuchans are currently Librans.

Dear Merlin,

    The tropic of Cancer is as far north as the Sun comes and the tropic of Capricorn is as far south as it goes. Does the Moon go farther north or south than the Sun?

        JAMES F. JACKSON

        CARLISLE, INDIANA

There is very much that is "tilted" in the solar system.

    The Moon's orbit is tilted five degrees and Earth's equator is tilted $23\frac{1}{2}$ degrees from the plane of the solar system. The Moon can therefore journey up to $28\frac{1}{2}$ degrees north *or* south of the celestial equator but only 5 degrees north or south of the Sun.

    When the Moon crosses the Sun's path en route from south to north or north to south) an eclipse will occur if the lunar phase is either new or full.

Dear Merlin,

If I were to go and stand directly on Earth's equator during the evening, which constellations would I see; the ones in the northern hemisphere, or the constellations in the southern hemisphere?

JULIE JONES

BERKELEY SPRINGS, WEST VIRGINIA

Residents of the equator are the only people on Earth who can see the entire northern sky *and* the entire southern sky across the year.

Merlin's good friend Santa Claus, when he is at home, sees only the northern sky.

Penguins, and other residents of Antarctica, see only the southern sky.

Dear Merlin,

In April 1986, my wife and I were in Chile to see Halley's comet. What we saw were two fuzzy comet-looking things. What were they?

J. O'BRIEN

DENVER, COLORADO

$\mathcal{M}$erlin is worried that in your trip to Chile you may not have seen comet Halley at all—if all you saw were two fuzzy things.

From Chile's vantage-point in the southern hemisphere, you should have seen four fuzzy things in the night sky.

If comet Halley was one of the fuzzy objects, then the second object you saw was probably the titanic globular cluster "Omega Centauri." Comet Halley passed close to it in the sky back in April 1986.

The other fuzzy objects are the two closest galaxies in the universe to your own Milky Way and they are found near each other in the sky. Ferdinand Magellan first wrote about them during the southern hemisphere interval of his around-the-world voyage.

Merlin just happened to be on that voyage and overheard Magellan describe the objects as "clouds." They are named in his honor: the larger (and brighter) of the two is called the "Large Magellanic Cloud," the other is called the "Small Magellanic Cloud."

Dear Merlin,

If the solar system is part of the Milky Way galaxy, what does it mean when people say they see the Milky Way in the night time sky?

MARJORIE HOLDER

NEW ORLEANS, LOUISIANA

If you have ever made blueberry pancakes you may have noticed that the berries poke out from the top and bottom. If you were the blueberry you would observe dense pancake in a circle around you and very little pancake above and below you.

Earth is as much a part of the disk of the Milky Way galaxy as your blueberry is part of its pancake.

When people observe the "Milky Way" they see a band of light (the collective light of billions of stars) that encircles Earth in the nighttime sky. Outside of this band, where there are fewer stars, the sky is considerably darker. The Milky Way's dense spiral disk is flatter than most pancakes, so this effect is quite pronounced.

Dear Merlin,

I once read a book that mentions in passing that Venus was once used by sailors to navigate during the day. What does Merlin have to say on this topic?

JOHN BRIDGES

GRAND PRAIRIE, TEXAS

The planet Venus can be seen easily in broad daylight if you have a small telescope or a good pair of binoculars and if you know where to look.

To navigate this way, however, is quite unnecessary in the daytime because the Sun (which you don't need a telescope to see) is available for this purpose.

If you need some excuse to navigate with Venus, then Venus is nearly always visible during the evening or morning twilight, which is when the Sun is below the horizon and the sky is too bright to see many stars. During this thirty-minute (or so) period Venus is alone, and visible with the unaided eye.

Looming brightly in the twilight sky, Venus has triggered untold phone calls to police departments by people claiming to see a UFO.

Dear Merlin,

My son and I are making a sundial/sun-calendar for a school science project. We noticed that the longest shadows did not occur on or about December 21—it looks like maybe January 5 or so had the longest shadow. There are many measurements that seem to indicate this. Could this be true or have we made a mistake gathering the data?

ROBERT J. FRITZ

BUDA, TEXAS

$\mathcal{M}$erlin suspects that you took all your readings at 12 noon.

If you adjust for how far away you are from the center of your time zone and if you correct for daylight saving time then the Sun will be at its highest point at 12 noon only four days of the year. All other days the Sun reaches its highest point a few minutes before noon or a few minutes after noon. These daily variations stem in part from the changing orbital speed of Earth during its elliptical journey around the Sun.

The longest shadows of the year *do* occur on December 21—they just don't occur at 12 noon. Your noon readings on January 5 caught the Sun before it reached its highest point (at a point lower than the noon sun on December 21) thus generating the longest noon shadow of the year.

The exact relationship for these noon variations has been endowed with the lofty name "the Equation of Time," and is graphically represented by a figure "8." Commercial sundials typically have this symbol etched somewhere on them while map and globe makers always seem to float it somewhere in the Pacific Ocean.

Dear Merlin,

How is it that the North Star can remain the North Star over the Earth's pole when the Earth is revolving around the Sun? Is it because the diameter of the circle caused by the Earth's revolution is negligible?

ADELE ROUNTREE
BEEVILLE, TEXAS

Yes. The 186-million-mile diameter of the Earth's orbit is quite negligible when compared with the three quadrillion miles to Polaris, the North Star.

Dear Merlin,

Is there a polestar for the South Pole the way Polaris is the polestar for the North Pole?

JENNIFER REGAN
FORT WORTH, TEXAS

T he South Pole has a star called "Sigma Octans" that is closer to the point in the sky directly above the South Pole than Polaris is in the north. But it is sixteen times dimmer so nobody pays attention to it.

Dear Merlin,

If satellites in orbit are stationed over some point on Earth, then why, on a dark night, do we see them streaking across the sky?

SONNY QUINTANILLA
SAN ANTONIO, TEXAS

Communications satellites typically have orbits large enough so that their period equals the rotation period of Earth. These are the "stationary" ones since they remain over a designated spot on the globe.

Other satellites (reconnaissance, military, geological, etc.) orbit Earth in less than two hours and are seen by reflected sunlight crossing the sky the first few hours after sunset, and the few hours that precede sunrise. They are easy to spot because they move fast, they don't blink, they don't have red and green wing lights, and they don't come in for a landing at your local airport.

Dear Merlin,

When will the next solar eclipse occur that will be visible to North Americans? (nontravelers, of course.)

G. K. STANTON
NORMAN, OKLAHOMA

*M*erlin does not know how old you are now, but you will be much older for the next eclipse that comes near Oklahoma. On August 12, 2045, no sooner, an O.K. solar eclipse will sweep across the entire United States from California to Florida.

If you don't mind leaving Norman, Oklahoma, then there is a total solar eclipse every few years somewhere on Earth.

VII

GRAVITY

$\mathcal{G}$ RAVITY, THE WEAKEST FORCE in nature, is the "glue" of the universe. All objects attract all other objects no matter how small or how far. The first understanding for this cosmic force was provided by Sir Isaac Newton. From his theory of gravity one can derive the reasons why stars are round, why planets orbit in ellipses, why objects that are tossed follow arched paths, why you are weightless when you are in orbit, why near satellites orbit quickly, why far satellites orbit slowly, why you weigh less on the Moon, and why you weigh more on Jupiter.

In the realm of the mundane, Newton's theory of gravity even describes why you weigh less after you have successfully dieted: you have less mass (the material content of your body); you and Earth are no longer attracted to each other as strongly as before; this force of attraction on Earth's surface is the definition of your weight; therefore you weigh less. When people want to lose weight, what they really want to do is reduce their mass; thus all weight watchers are really mass watchers.

Dear Merlin,

What is the real story about Isaac Newton and the apple? I've heard a lot of contradictory tales.

B. A. MILSAP

LONG ISLAND, NEW YORK

$\mathcal{M}$erlin recently had a chat with Sir Isaac Newton about his discovery of the laws of gravity. It was the year 1710 in the backyard of his Lincolnshire home in the English countryside.

Sir Isaac explained, "I went to sit under a tree to contemplate the cosmos when I saw an apple fall from the branch of a distant tree. The Moon happened to be visible in the sky the moment the apple fell." He continued, "I then postulated that the same laws of gravity that attracted the apple to Earth also kept the Moon in orbit."

Merlin had to interject, "You mean you were never really hit in the head with an apple?"

Sir Isaac responded, "Merlin, if I had been hit in the head with an apple while contemplating the cosmos, the only revelation that would have come to me would be to move to a different tree!"

Dear Merlin,

Since centrifugal forces oppose the pull of gravity, how much less do objects weigh due to the Earth's rotation?

B. B. RININGER

LONG BEACH, CALIFORNIA

*O*n the equator, where centrifugal forces are greatest, a 150-pound person will be a slender 149 pounds 14 ounces. Polar dwellers receive no such benefits from the rotation of the Earth and so maintain their "true" gravitational weight, although this fact does not explain why Santa is so chubby.

Dear Merlin,

How fast does the pull of the Earth's gravity decrease as one leaves the surface of the Earth?

DAVID MILLER
PITTSBURG, CALIFORNIA

*M*erlin once asked Sir Isaac Newton, a good friend, about the force of gravity. Isaac responded: "If spheres be however dissimilar . . . I say that the whole force with which one of these spheres attracts the other will be inversely proportional to the square of the distance of the centers."

He later published this idea in 1686 as Proposition LXXVI in *The Principia*, Book I.

If we apply Newton's law of gravity to you and Earth (where you are the "dissimilar" sphere) we find that every time you triple your distance from Earth's center the force drops to one ninth of its previous value. If you quadruple your distance, the force drops to one sixteenth, etc.

But if you are just going up and down elevators, your distance to Earth's center doesn't change very much so the variation in the force of gravity is correspondingly small.

Dear Merlin,

Could you throw a rock from a spaceship and get it into Earth orbit?

SEAN BISHOP
CONCORD, CALIFORNIA

That all depends on where your ship is when you throw the rock.

If your spaceship is parked on the launchpad then all you need do is toss the rock at 18,000 miles per hour toward Earth's horizon. It will immediately enter low Earth orbit.

If your spaceship is already in orbit above Earth then just roll down the window and place the rock outside. It will remain near the window in orbit around Earth with the ship.

If your spaceship is traveling through a distant part of the universe, en route to another galaxy, then you may no longer be concerned about throwing rocks around Earth.

Dear Merlin,

If the astronauts in the Space Shuttle are still within the
gravitational pull of Earth when they are in orbit then how can
they be weightless?

JONATHAN LEVY
ORLANDO, FLORIDA

*A*ny object that finds itself falling freely toward the
surface of Earth will be weightless.

If a freely falling astronaut, a few hundred miles above
the ground, also happens to have a sideways motion of
18,000 miles per hour (kindly provided by the Space
Shuttle) then the astronaut will fall toward Earth at the
same rate that the round surface of Earth curves away.
This condition is commonly called an "orbit."

Dear Merlin,

If an object in orbit around Earth is considered to be in "free fall" then why is it not accelerating and eventually overcoming the fall-away of the round surface of its host?

JOHN SIMBLES

SACRAMENTO, CALIFORNIA

In physics, an acceleration can be an increase in speed, a decrease in speed, and/or a change in the direction of motion. An object in circular orbit around Earth is accelerated by Earth's gravity in a way that changes its direction of motion but not its speed. Merlin's good friend Sir Isaac Newton first showed this effect in his famous treatise on mechanics, *The Principia*, published in 1686.

Dear Merlin,

What does it mean when a satellite has been lifted to a "geosynchronous" orbit?

PAUL KRONKOSKY

HOUSTON, TEXAS

 T he farther above Earth's surface a satellite orbits, the longer it takes to complete an orbit. At the altitude of about 23,000 miles the satellite takes 23 hours and 56 minutes to orbit Earth. Earth takes 23 hours and 56 minutes to rotate. This arrangement permits a satellite to "hover" over a chosen spot of the Earth's surface.

Communications satellites are routinely given this type of orbit.

Dear Merlin,

Why is the 23,000-mile altitude for a satellite in geosynchronous orbit so special? Why cannot satellites be at say 30,000 miles and have a larger orbital speed that will still keep them in geosynchronous orbit?

T. WIGNARAJAH
TAMUNING, GUAM

For every altitude above Earth's surface there is only one speed that will sustain a circular orbit for an artificial satellite.

If you tried to speed up a satellite it will jump to a higher orbit and travel slower than it did before. If you try to slow down a satellite it will fall to a lower orbit and travel faster than it did before.

Satellites in a very low orbit travel about 18,000 miles per hour and complete an orbit in about 90 minutes. Satellites in geosynchronous orbit travel about 7,000 miles per hour and, of course, complete their orbit in 23 hours 56 minutes, the rotation period of Earth.

Dear Merlin,

What is the "tidal force"? Is it some new kind of force or is it really related to oceanic tides?

BETH ANN MARTIN
PITTSBURGH, PENNSYLVANIA

Tidal forces are produced by the force of gravity.

A planet (or any object) will feel a tidal force if one side of it is closer to a source of gravity than the other side. Gravity will then pull stronger on the nearer side than the farther side and the planet will feel "stretched" toward the source of gravity. This stretching effect is known as the "tidal force."

The tidal forces from the Moon and Sun are not strong enough to rip apart the solid Earth, but the liquid oceans bulge quite readily to form the common "tides."

Dear Merlin,

I have often wondered and marveled at the ability of unmanned spacecraft to get so close to other planets. How are they able to get so close to those far away objects without running out of fuel? What makes them so accurate?

STEVE FISHER

COSTA MESA, CALIFORNIA

$\mathcal{M}$ost of the time, spacecraft do not burn fuel at all. They just "coast" to their destination after the last rockets are fired. There is no air friction or road friction in space so that once you are set in motion, gravity is the only force that affects your trajectory.

If you want to catch a planet, you don't aim straight for where it is, because by the time you get there, the planet will be somewhere else in its orbit. After a careful study of the planet's orbit and after you compensate for Earth's orbital speed and position (assuming you leave from Earth), just aim to where you expect the planet to be when you get to its orbit.

If you now make small adjustments in your trajectory (remembering that as you approach, the gravitational influence of the target planet is growing) you can choose to crash-land your craft by aiming straight at the planet. If that doesn't please you then you can "fly-by" it by plotting your trajectory slightly to the planet's side. Depending on the details of your adjustments, the gravity will either capture you in orbit, or it will swing you around the planet and jettison you back into space in another direction.

Dear Merlin,

I once read that the planetary probe Pioneer 10 *was accelerated to the velocity of escape for the solar system by the gravitational field of Jupiter. According to my understanding of Newton's laws,* Pioneer 10, *after fuel burnout, has constant total energy—kinetic plus potential. If* Pioneer 10 *did not have escape energy at the moment of burnout, the gravitational field of Jupiter could not give it to it. Jupiter's field merely causes the ratio of kinetic to potential energy to change with time. Can you point out the fallacy in this reasoning?*

> PHILIP HAYWARD
> GAITHERSBURG, MARYLAND

Yes.

Depending on the exact trajectory of *Pioneer 10* it can acquire extra speed that is up to, but does not exceed, the orbital speed of Jupiter. To accomplish this it actually *takes away* momentum from the planet Jupiter in its fly-by.

Jupiter, which is more massive than all other planets combined, has plenty of momentum to forfeit to the tiny planetary probes. After each fly-by, Jupiter remains in its orbit essentially unaffected.

Dear Merlin,

Would the spacecraft Pioneer 10 be able to detect a planet beyond the orbit of Pluto by gravitational influence if the possible planet was on the opposite side of the Sun when Pioneer 10 crossed its orbit?

DANIEL COLEMAN
HOUSTON, TEXAS

*N*o.

Dear Merlin,

What is the fastest object that humans have ever sent into space?

MARYLIN RIVERA

SAN ANTONIO, TEXAS

The American planetary probes *Pioneer 10* and *11* were launched in 1972 and 1973. They each achieved a speed in excess of 40,000 miles per hour (12 miles per second) upon flying by the planet Jupiter. The follow-up probes *Voyager 1* and *2* duplicated this dynamical encounter with Jupiter in 1979.

These four spacecraft have the unique distinction of being the only spacecraft to travel fast enough to completely escape the gravitational pull of the solar system. They will enter the unvisited realm of interstellar space.

The space speed record, however, must go to the U.S.-German solar probe *Helios 2*, launched in 1976. It was clocked at nearly 150,000 miles per hour (42 miles per second). Note that this is only one fiftieth of one percent of the speed of light.

Dear Merlin,

Does the Sun's gravity end at Pluto?

G. THORNDIKE
TROY, NEW YORK

If the Sun's gravity ended at Pluto then we would have lost Neptune back in 1979. This was when Pluto, with its very flattened orbit, crossed the orbit of Neptune to become the eighth planet from the Sun.

There is also a region of solar system debris that extends half-way to the nearest stars. It is believed to be the Sun's primary source of comets.

Regardless, Isaac Newton's laws of gravity dictate that the Sun's gravity will get weaker and weaker the farther out you go—yet will never reach zero.

VIII

S T A R S

STARS LEAD TRANQUIL LIVES yet undergo rather traumatic deaths.

Their energy is produced by the thermonuclear fusion of elements. Stars with a fraction of the Sun's mass use their available "fuel" quite efficiently. Some of these stars are expected to outlive the universe.

Stars that are as massive as the Sun will eventually exhaust their "fuel" and swell up a thousandfold like an enormous beach ball. Ultimately, the expanded outer shell of star material lifts away to lay bare a small dense white-hot core—the "white dwarf."

Stars several times more massive than the Sun will also swell to a thousand times larger but their "fuel" consumption occurs rapidly and exotically. The core of these stars is an alchemist's dream. They convert hydrogen to helium, helium to carbon, carbon to nitrogen, and oxygen and silicon . . . all the way to the element iron. At iron the star can convert no longer. It collapses under its own weight and recoils with an enormous energy—the "supernova."

Stars many times more massive than the Sun behave similarly to those that are just several times more massive than the Sun. The difference occurs when the very high mass star does not recoil. It collapses continuously to form a gravitational hole in space that does not emit light—the infamous "black hole."

Dear Merlin,

Is there a chance that another star will one day collide with the Sun?

KATHERYN BECK
OLYMPIA, WASHINGTON

Yes.

But you should know that if there were just four snails randomly carousing throughout the continental United States then it is more likely that two of them will accidentally bump into each other than it is for another star to collide with the Sun.

Dear Merlin,
How big is the biggest star?
Paul Milstead
Lawrence, Kansas

The class of stars called "super giants" heads the list of biggest stars. Most prominent among them is Betelgeuse, in the constellation Orion. It is an irregularly pulsating red super giant that—if it were put in place of the Sun—would vary in size from the *orbit* of Mars to the *orbit* of Jupiter.

The charred remains of Earth would be found orbiting deep within Betelgeuse's gaseous interior.

Dear Merlin,

Are red giants red because they are hotter than the Sun?

EVELYN CHAMBERS
PORTLAND, OREGON

No, they are red because they are cooler than the Sun.

On canvas with paint
In the Artist's school
It's red that is hot
And blue that is cool.

But in science we show
As the heat gets higher
That a star will glow red
Like the coals of a fire.

Raise the heat some more
And what is in sight?
It's no longer red
It has turned bright white.

Yet the hottest of all,
Merlin says unto you,
Is neither white nor red
When the star has turned blue.

Dear Merlin,

My question is about binary stars. But first correct me if I'm wrong. 1) All stars have mass. 2) All mass has at least some gravitational pull on other objects. If this is true then how can binary stars exist? Would they not draw themselves together?

D. L. F.
<small>GREENFIELD, WISCONSIN</small>

If binary stars just sat there in space then their mutual gravity would certainly draw them together.

But when an object is in motion around any other object the force of gravity serves primarily to bend the direction of motion into a closed loop. We now have what is commonly called an "orbit" and we have prevented an unpleasant cosmic collision.

Dear Merlin,

I have a question which I'm embarrassed to ask, but I want to know the answer as the basis for a possible short story.

If our star, the Sun, were a double star, in other words if we on planet Earth were revolving around two suns of the same magnitude as our single Sun, would objects and beings on Earth cast a double shadow? Two shadows for earthlings instead of one?

CATHERINE SARNELLI
LEVITTOWN, NEW YORK

Yes.

Dear Merlin,

What is the difference between a protostar and a brown dwarf?

GORDON N. COOMER
BASTROP, TEXAS

A protostar is a large gas cloud that is collapsing and getting hotter in an attempt to ignite thermonuclear fusion at its center. If it succeeds . . . a star is born.

If the protostar is not massive enough the thermonuclear fusion never begins. What's left is a ball of hot gas that cools and shrinks. That is what is commonly called a "brown dwarf."

If a brown dwarf that orbits another star is given sufficient time to cool—so that it shines primarily by reflected light from the star it orbits—then it may be reclassified as a planet.

Dear Merlin,

What is the most number of stars that have been found in one star system?

ERIN FRENCH
LANSING, MICHIGAN

$\mathcal{M}$erlin's good friend, Sir John Herschel, commented over a century ago, "The noble globular cluster Omega Centauri is beyond all comparison the richest and largest object of its kind in the heavens. The stars are literally innumerable. . . ."

We now know this cluster to contain over a million stars that are 25,000 times more densely packed than in the solar neighborhood.

Dear Merlin,

What happens to any planets, moons, or other stellar objects in the immediate area of a star that becomes a supernova?

BARBARA GRAYSON

FORT WORTH, TEXAS

*A*ll nearby celestial objects will be bathed in heavy doses of radiation of all wavelengths—especially the higher energy wavelengths that include ultraviolet, X rays, and gamma rays. The shock wave brought on by this radiation will fragment and dissipate all interstellar gas clouds in the area. Planets and moons, after getting their surfaces vaporized, will be ejected into interstellar space. The varied chemical composition of the exploding star will contribute to the interstellar enrichments of elements heavier than the primordial hydrogen and helium. Future generations of stars and planets and moons will then condense out of this enriched interstellar soup to form celestial objects and entities of increasing chemical diversity such as life.

Dear Merlin,

If a nearby star were to become a supernova, I know we would see it but would be hear it?

GWEN BEAM

COLUMBIA, MARYLAND

In space, not only can no one hear you scream, no one can hear you explode either.

Unlike light, sound requires a material medium in which to move (air, liquid, solid). Interstellar space isn't completely empty, but it is empty enough to render mute a cataclysmic explosion such as a supernova.

Dear Merlin,

I once read about a star in the universe that spins five hundred times per second. It must have an immense amount of mass and gravity to balance the centrifugal force. What creates such a spin in the first place?

TOM DIX

OCEANSIDE, CALIFORNIA

*I*f you suck a long strand of gently swaying spaghetti into your mouth, it will wiggle more and more vigorously until the last three inches when you are guaranteed to have it flap into your face.

If the arms of a figure skater who spins slowly are drawn inward, the figure skater will spin faster and faster.

If a rotating gas cloud collapses to form a star, it rotates faster and faster until it breaks apart unless the cloud's gravity is strong enough to keep it contained.

All three scenes are examples of a general principle in physics where a shrinking and rotating object spins faster and faster. It is called the "conservation of angular momentum." It works on your dinner plate and it works in the cosmos.

P.S. Figure skaters do not spin fast enough to break apart.

IX

LIGHT & TELESCOPES

LIGHT & TELESCOPES

*A*STRONOMERS ARE THE WORLD'S EXPERTS when it comes to light. The entire science is based on the collection and analysis of light from objects that are not on Earth. If an object to be studied were on Earth you could walk up to it, pick it up, and play with it in any way you please. Astronomers have no such luxury. Astronomers can't reach out and grab a piece of star for laboratory analysis. Astronomers can't reach out and tilt a galaxy to get a different angle of view. Astronomers can't watch the birth *and* death of a star because stars live a million (up to a trillion) times longer than humans. In short, astronomers can't control any object or phenomenon they observe. In this way, astronomy is the most humble of all disciplines. There is, however, an enormous amount of information that can be extracted from the light of celestial objects. Indeed this was how information about the origin and expansion of the universe was revealed.

There are few professions where bigger always equals better. The construction of telescopes for astronomers is one of them. Like collecting raindrops with a bucket rather than a thimble, larger telescopes, with their greater surface area, collect more light than smaller telescopes. They thus enable astronomers to probe dimmer and farther reaches of the universe.

Dear Merlin,

If there are rainbows from the Sun and moonbows from the Moon are there starbows too?

PETER DINSMORE

HELENA, MONTANA

Rainbows tend to appear when it is raining at the same time the Sun is shining.

Moonbows tend to appear when you look at the Moon through a hazy, semitransparent cloud of tiny ice particles.

If ordinary stars were bright enough (anywhere from a million to a billion times brighter) then they would make their own version of these bows and we could happily call them "starbows."

In this naming scheme, however, we should rename rainbows as "sunbows."

Dear Merlin,
Does UV light and visible light travel faster than infrared light?
Tommy Jo Mitchel
Dallas, Texas

N₀.

All electromagnetic radiation travels at exactly the same speed—the speed of light. This includes, in order of increasing energy:

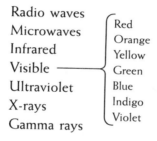

Radio waves
Microwaves
Infrared
Visible ———
Ultraviolet
X-rays
Gamma rays

Red
Orange
Yellow
Green
Blue
Indigo
Violet

Dear Merlin,

I would like to know what happens to light as it goes through empty space. Does it decay? Does it slow down? If so, what effect would this have on its appearance? Would it change color or wavelength?

BRUCE PHILLIPS
QUINLAN, TEXAS

Light gets dimmer in accordance with the well-established "inverse square law." For example, if you move three times as far away from an object it looks nine times dimmer.

If the object that emitted the light is in motion—as is true for galaxies in our expanding universe—we have a shift in the wavelength (color) of the light. The spectrum is red-shifted for receding objects and blue-shifted for approaching objects. The pioneering research on wavelength shifts was conducted using sound waves in the mid-nineteenth century by the Austrian physicist Christian Johann Doppler. Red shifts and blue shifts are now called "Doppler Shifts" in his honor.

The speed of light remains unchanged and is the same for all observers.

Dear Merlin,

    How many photons per second does the Sun give off?

    CHRISTOPHER SHAW

    PHOENIX, ARIZONA

*M*erlin hasn't counted lately.

When you are as busy as Merlin, you do not have time to count every photon. Fortunately, you can estimate the answer by dividing the total luminosity of the Sun by the average energy per photon emitted. These are well-researched and documented quantities.

After some computing on Merlin's abacus, Merlin derives approximately 10^{45} photons per second.

Dear Merlin,
  What is a light-year?
  ELIZABETH NORRIS
  PHILADELPHIA, PENNSYLVANIA

A light-year is not a year with fewer calories, nor is it an interval of time. A light-year is a *distance* equal to about 5,800,000,000,000 miles. A beam of light in space will travel this far during one Earth year.

It is a convenient yardstick for describing cosmic distances. For example, it is somewhat easier to say that the nearest star to the Sun, Proxima Centuri, is 4.1 light-years away rather than 23,800,000,000,000 miles.

Dear Merlin,

I'm confused about the terminology surrounding detection ability. What is the dimmest magnitude visible with my naked eye? What is the dimmest magnitude visible with my 7 × 50 millimeter binoculars? What is the dimmest magnitude visible in my 60 millimeter refractor?

JAMES O. ROBERSON
WILLIAMSTON, NORTH CAROLINA

On the inverted logarithmic "magnitude" scale used by astronomers to identify brightness, the Sun is −26, the full moon is −12, Venus (at its brightest) is about −4, and Polaris the North Star is +2.

If you have good eyes you can see down to 6th magnitude with the unaided eye. (Merlin prefers "unaided" rather than "naked.") This will get you about five thousand stars, six planets, the Sun, the Moon, and Merlin's home—the Andromeda galaxy.

Your 7 × 50mm binoculars will permit you to see down to 11th magnitude. You will promptly jump from five thousand to five million stars. Nebulae, star clusters, and many galaxies are added to your discovery list.

Your 60mm refractor will not see much dimmer than the 7 × 50mm binoculars but the image will be considerably more magnified. Note that "magnification" only enlarges the image already produced by the main lens (or mirror).

On Mauna Kea, Hawaii, the 10-meter Keck telescope

can detect objects to about 28th magnitude. We now get to amend our list with all known quasars, and billions of galaxies. Because bigger telescopes detect dimmer objects, the largest telescope in the world defines the known visible "edge" of the universe.

Dear Merlin,

When we see those pictures of X-ray sources and radio galaxies do scientists just read certain bands of the light or do they send the light through some sort of special prism?

MARJIE LEONG

DALLAS, TEXAS

*A*stronomers are often categorized by the type of light they study. There are radio astronomers and microwave astronomers and infrared astronomers and gamma ray astronomers. Indeed, astronomers are all over the spectrum.

Each part of the spectrum requires its own specially designed telescopes, detectors, and filters to produce an image of the objects under study. A common and useful band of light for optical astronomers is the "B" band, which corresponds to blue wavelengths of light. For radio astronomers it is the famous 21-centimeter wavelength emitted by the hydrogen atom.

When telescopes are "tuned" to these and other bands their detectors generate intensity maps that recreate an image of what the object looks like in the chosen wavelength of light.

Spectroscopists pass light through expensive versions of prisms to analyze all wavelengths at once, but this method does not produce an image.

Dear Merlin,

Can you tell me, now that astronomers are using the entire electromagnetic spectrum, what future developments along these lines can we expect?

ALBERT J. CLEMENT
MADISON, NEW JERSEY

$\mathcal{M}$erlin does get around the cosmos, but Merlin has never claimed to know the future.

Briefly, current research includes the mapping of galactic structure with unprecedented resolution from specially designed arrays of radio telescopes.

Microwave telescopes are used to map the cosmic background radiation, as well as dense gas clouds where star formation is expected to occur. Ground-based and orbiting infrared telescopes are used to locate star-forming regions that are enshrouded by these dense gas clouds. They can also detect planets in the process of condensing in the vicinity of a newly formed star.

Optical telescopes are still the workhorse of the science. A frontier in this realm is the 94-inch Hubble orbiting "Space Telescope." Its position above the Earth's lower atmosphere permits extremely accurate measurements of extremely distant galaxies, as well as stellar proper motions, which are likely to indicate the presence of planets. It can also measure stellar parallaxes to an unprecedented accuracy, which continues to improve the cosmic distance ladder.

Ultraviolet, X-ray, and gamma-ray wavelengths are used extensively in the study of "active" galaxies. This broad category of celestial wonder includes quasars, Seyfert galaxies, starburst galaxies, and any galaxy where a supermassive black hole may be the central engine that is responsible for the high-energy emission.

Other frontiers involve nonelectromagnetic signals. Neutrinos are predicted to be emitted during stellar collapse. In 1987, for the first time, neutrino detectors successfully measured the emission of high-energy neutrinos that resulted from the well-publicized "Supernova 1987a" in the Large Magellanic Cloud.

Lastly, the scientific community eagerly awaits the detection of gravity waves (predicted in Einstein's General Theory of Relativity) from objects such as supernovae, exploding galactic nuclei, and stellar gravitational collapse.

Dear Merlin,
   What is the largest telescope in the world?
      ANDY SIMPSON
      ST. LOUIS, MISSOURI

 T he largest telescope in the world is the nonsteerable radio telescope dish in Arecibo, Puerto Rico. It is over three football fields in diameter and is built in a natural crater on the Earth's surface. The telescope can be "aimed" by shifting the receivers that are suspended high above the central area of the dish.

The largest optical telescopes are the twin 10-meter (400-inch) Keck telescopes with segmented mirrors on Mauna Kea, Hawaii.

The largest single-mirror optical telescope is the 6-meter (236-inch) reflecting telescope on Mount Semirodriki, Caucasus.

The largest refracting telescope is the 40-inch at Yerkes Observatory, Williams Bay, Wisconsin.

The Very Large Array (VLA) radio telescope, fifty miles west of Socorro in the plains of San Augustin, New Mexico, is a Y-shaped rail system where each arm is about thirteen miles long. Along these arms are twenty-seven mobile receiving dishes, each 82 feet in diameter. They are simultaneously steerable by a centralized computer. The data collected from all dishes are summed together to create the effect of a single radio telescope dish that is more than thirty-five miles in diameter.

Dear Merlin,

If the North and South Poles get six months of darkness at one stretch then why don't astronomers build telescopes in these regions?

DELORES PAGE

AUSTIN, TEXAS

The most stable condition for observing at a domed telescope occurs when the air temperature inside the dome equals the air temperature outside the dome. Observational astronomers tend to be a hardy bunch. (Winter temperatures at many mountain telescope sites commonly dip below zero degrees Fahrenheit.) Merlin firmly believes, however, that none of them are interested in risking their life at a telescope dome whose air temperature has equaled the winter polar temperatures of 100 degrees *below zero.*

An easy solution would be to conduct observations remotely from a toasty warm observing room thousands of miles away, but another factor involves the northern and southern aurorae (aurora borealis and aurora australis). During sunspot maximum these colorful atmospheric displays can light up the sky for hundreds of consecutive nights thus beautifully interfering with observations.

Dear Merlin,

How effective is the Hubble Space Telescope be for Earth viewing?

V. W. ROBERTS
HOUSTON, TEXAS

There are already satellites whose sole purpose is to photograph Earth. The Hubble Space Telescope is more powerful than all of them—an ideal peeping tom. On a clear day, it could easily observe the car you drive.

Astronomers, however, are not interested in spending billions of dollars to raise a telescope three hundred miles into space just to look back down through the atmosphere at Earth.

Dear Merlin,

> *Can the Space Telescope see some stars as a "disk"?*
> DOMINICK DESIMONE
> STATEN ISLAND, NEW YORK

*N*o.

The 94-inch Hubble Space Telescope can see all sun-size disks out to about one light-year. But there are none.

It can also see the disk of all red giants out to about two hundred light-years. But there are none.

Dear Merlin,

I have a question about the speed of light. I've been told that the closer an object comes to attaining light speed, the larger its mass becomes. What is the notion behind this?

BARRY BRYANT

HURST, TEXAS

*A*n object's energy of motion—its kinetic energy—is related to its speed and its mass. When an object is in motion we normally increase its kinetic energy by increasing its speed.

At speeds that are an appreciable fraction of the speed of light we find that we can still increase the kinetic energy of the object but the mass is also involved.

The speed of light (186,282 miles per second) has been experimentally shown to be a physical limit to all speeds. Nature appears to be well aware of this comic speed limit and responds to an increase in an object's energy of motion by an increase in the object's mass as well.

Merlin's good friend Albert Einstein first described this phenomenon in 1905 as a consequence of his Special Theory of Relativity.

X

GALAXIES GALORE

*I*F, WITH THE AID OF A TELESCOPE, you look beyond the nighttime stars of the Milky Way you will see innumerable fuzzy specks called galaxies. These are the distant gravitational collections of billions of stars that form the basic constituent of the visible universe. To the residents of these galaxies, the Milky Way and Andromeda look like fuzzy specks in their nighttime sky.

Galaxies are found isolated, in pairs, in quintets, in groups, in clusters, and in super clusters. Galaxies are found that are hundreds of times more massive down to ten thousand times less massive than the Milky Way.

In Merlin's opinion, galaxies are the most beautiful of all celestial wonders that bespeck the space of the universe.

Dear Merlin,

What is the farthest the naked eye can see in the nighttime sky?
ARNOLD WEBB
LOS ANGELES, CALIFORNIA

On a smoggy evening in your home city of Los
Angeles, Merlin estimates about two hundred yards.
Under favorable conditions, however, one can expect to
see two million light-years away to Merlin's home, the
Andromeda galaxy. It is the farthest object visible to the
unaided eye. The galaxy's three hundred billion stars
appear as a fuzzy patch in the constellation Andromeda.

Every other star in the sky is part of your own galaxy,
the Milky Way, and is within a few thousand light-years
of the Sun.

Dear Merlin,

If the Andromeda galaxy is over two million light-years away then when I look up into the sky with binoculars, do I see the light it emits now or is the light two million light-years old?

RON BURMAN

CHICAGO, ILLINOIS

You see Andromeda not as it is and where it is, but as it was and where it was over two million years ago.

The photons you see left Andromeda at the dawn of the Quaternary Period of Earth's geologic history.

Dear Merlin,

What is the closest galaxy to our own, Andromeda, or Large Magellanic Cloud?

AMY FREDERICKS
CHEVY CHASE, MARYLAND

*A*t two million light-years, Andromeda is over ten times farther than the two "satellite" galaxies to the Milky Way, the Large and Small Magellanic Clouds.

Dear Merlin,

   How many types of galaxies are there?
   KAREN SCHINDLER
   VICTORIA, BRITISH COLUMBIA

There are three main types: elliptical galaxies, spiral galaxies (like your own Milky Way), and irregular galaxies.

Quasars, Seyfert galaxies, radio galaxies, starburst galaxies, cD galaxies, lenticular galaxies, and dwarf galaxies are all exotic versions and combinations of those three broad categories.

Dear Merlin,

What are the largest galaxies? Are they the elliptical or the spiral types?

ED CRASTON
SEATTLE, WASHINGTON

The gas-rich, star-forming spiral galaxies may be pretty, but elliptical galaxies hold all the records. The largest, the brightest, the smallest, and the dimmest known galaxies are all some variety of elliptical galaxy.

The cD galaxies tend to be the largest of the elliptical family. They are found most often looming near the centers of crowded galaxy clusters. Many have multiple nuclei, which lends the suspicion that whole galaxies have been cannibalized.

A cD galaxy can be over ten times the mass of the entire Milky Way.

Dear Merlin,

How many galaxies are there in the universe?

AMELIA MORAIDA

AUSTIN, TEXAS

Billions and billions and billions.

The latest estimates have ranged anywhere from ten billion to one hundred billion galaxies. The most numerous constituent of the universe may be the hard-to-detect dwarf galaxies that are small and dim. Data from the Local Group of galaxies suggest that there may be more dwarfs than all other galaxies combined.

Dear Merlin,

If I understand correctly, the Milky Way in the night sky is what we can see of the bulk of the spiral galaxy that we and the Sun are part of. Where are all of the brighter stars, most of which seem far away from the Milky Way from our perspective?

JOHN HILL

LAS VEGAS, NEW MEXICO

*E*arth and the Sun and the entire solar system are embedded in a flattened spiral disk of hundreds of billions of stars called the Milky Way galaxy. The band across the night sky is the collected light from the stars of the Milky Way's disk.

Every other star you see is also part of the disk, but since we are embedded deep within it, we see stars all around us in the sky.

Dear Merlin,

I understand that the globular clusters are arranged spherically around the center of our galaxy. When the Milky Way originally formed, how was it possible for the globular clusters to keep their original spherical arrangement while the rest of the galaxy collapsed to form a disk?

JOHN APPLEDOORN
SAVANNAH, GEORGIA

*W*hen the initial spherical gas cloud of the Milky Way started to form stars (over ten billion years ago) it used only a small fraction of its gas to make the globular clusters. The rest of the gas collapsed from above and below the midplane of the galaxy to make the disk—when a gas cloud collides with another gas cloud they are more likely to fuse together (like two hot marshmallows) than to pass through each other.

Any globular cluster that encountered the disk would easily bulldoze its way through the gas clouds to continue its orbit through to the other side of the disk.

The resulting picture of the Milky Way is a flattened gas-rich disk of old and new stars with a spherical gas-empty "halo" of very old globular clusters.

Dear Merlin,

I have often been thrilled to gaze in the direction of Sagittarius and Scorpius and know that I was looking at the center of our galaxy. Similarly, I know that as I look toward the southern part of Auriga, also in the Milky Way, I am looking toward the outer edge of the galaxy, opposite the center. Can you tell me which direction (toward which constellation) we (the solar system) are headed as we revolve around the center of the galaxy? Also, which constellation is in the direction that the galaxy or local cluster of galaxies is headed?

NICK JACKSON

JUNCTION, TEXAS

The rotation of your galaxy carries the Sun and environs at about 130 miles per second in the direction of the constellation Vela, the Sail. The Milky Way galaxy, the Andromeda galaxy, and the entire Local Group of galaxies are falling at about 250 miles per second on a collision course with the center of a large cluster of galaxies in the constellation Virgo. Such is the layout of this cosmic ballet.

There is no need to worry about all this, however, because we don't expect a major intergalactic collision for several billion years.

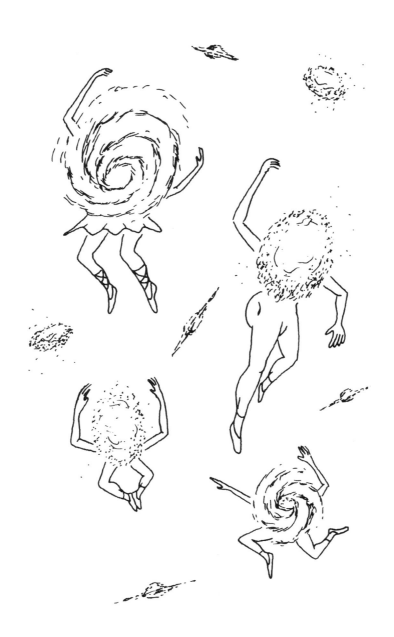

Dear Merlin,

How have astronomers been able to measure the dimensions of the Milky Way? I can see that we on Earth can observe the "right" and "left" sides of the galaxy by taking measurements at two opposite points in our revolution. But what about its height? How do we know it's a platter with a bulge in the center?

GORDON E. CASTANZA

BEIJING, CHINA

*A*stronomers have yet to figure out how to live for 250 million years. It takes about that long for the Sun to revolve around the galaxy center. "Right" and "left" measurements are therefore out of the question.

The dimensions of the Milky Way are deduced from an involved set of indirect methods, most of which use the 21cm radio wavelength (emitted by the hydrogen atom) and the infrared wavelengths (emitted by heated gas) to map the presence of gas clouds. These long wavelengths of light easily penetrate the gas and dust clouds that prevent us from seeing the galactic structure directly.

From the rotation of the galaxy we find gas clouds that move with different speeds relative to the solar system. By carefully measuring where they are and how fast they are moving we reconstruct the solar system's location among the spiral arm branches of the Milky Way.

The height can be inferred from the transparency of the galaxy's disk as you look above and below it through the gasless halo (home of the globular clusters) out to the rest of the universe.

The location and extent of the bulge is plainly visible in wide field infrared photography across the galactic center.

Dear Merlin,

I am forty-two years old because I've encircled the Sun forty-two times. How old is the Sun for having circled the center of the galaxy?

RICHARD LAWRENCE
INDIANAPOLIS, INDIANA

With a little help from Earth, you have orbited the Sun forty-two times at a speed of 18 miles per second. The Sun has an orbital speed of about 130 miles per second. Even at this blazing speed it has only encircled the center of the Milky Way galaxy about twenty times in the five-billion-year history of the solar system.

Officially, then, the Sun is a youthful twenty *galaxian* years old.

Dear Merlin,

Do galaxies ever collide? What would they look like if they did?

DAVID BALDWIN
CINCINNATI, OHIO

The Milky Way is in no imminent danger but galaxies that are members of rich clusters appear to collide quite often. The merging gravity fields create havoc among stellar orbits—casting stars to and fro. The ongoing scene is the galactic equivalent of a train wreck. Space is so vast that the stars themselves will miss each other, but the larger interstellar gas clouds will collide, undergo enhanced star formation, and get left behind after the collision.

Dear Merlin,

Why is it that galaxies do not lose their relative positions in the universe even though we are all traveling through space?

JOHN H. (BUD) HELK

BREMENTON, WASHINGTON

Galaxies do lose their relative positions.

You can actually watch this happen but you probably have better things to do. If you lived for a hundred million years you would get to see many galaxies move from their current position by an amount equal to their diameter.

Dear Merlin,

Can a person cross our galaxy in a spaceship during one human life span?

PRINCE RICKER

AUSTIN, TEXAS

$\mathcal{U}$nlike Captain Kirk and the Starship Enterprise— where they routinely crossed the galaxy during television commercials—a real person in a real spaceship is restricted by the speed of light. Light, the fastest thing we know, requires 100,000 years to cross the Milky Way galaxy. Humans have yet to live that long.

In 1905 Merlin's good friend Albert Einstein introduced the Special Theory of Relativity, which predicts that time will tick slower and slower the faster you travel. Were you to embark on such an adventure you could conceivably age as little as you wish, depending, of course, on your exact speed. The problem arises when you wish to return to Earth (assuming that's where you started). Earth will have moved several hundred thousand years into the future and everyone will have forgotten about you.

Dear Merlin,

I'm confused about Hubble's law and the relationship between velocity and the distance to a galaxy. Is that relationship assumed, or has it been established by independent means?

DAVID DETWILER
HOUSTON, TEXAS

*A*nother one of Merlin's good friends, the American astronomer Edwin Hubble, first established the velocity-distance relation for galaxies in 1929.

Earlier data collected by V. M. Slipher in combination with Hubble's data showed most galaxies to be receding from the Milky Way. It was the first indication that we were part of an expanding universe.

The velocity-distance relation was established using the highly luminous "Cepheid" variable stars that were visible in nearby galaxies. Their period of light variation is tightly correlated with their luminosity. With the help of some simple equations, a distance can be deduced for the galaxy in question. Today, many more distance indicators have been used to derive a more accurate velocity-distance relation than Hubble's original results.

If a galaxy is too far away for any of the distance indicators to be useful then the Hubble relation is assumed to be valid and the distance is inferred purely from the galaxy's velocity of recession.

Dear Merlin,

Since nothing travels faster than light, how can there be redshifts greater than one?

Joe Stickler

Valley City, North Dakota

$\mathcal{R}$edshifts in the universe simply measure the effective "stretching" of wavelengths of light emitted by galaxies moving away from us. A fast enough galaxy can stretch wavelengths very much more than the original value— thus producing redshifts greater than one.

Confusion arises because an approximate formula (*redshift* $\approx v/c$: the velocity of the galaxy divided by the velocity of light) is used for nearby galaxies instead of a more complicated exact formula, which uses the principles of Einstein's Special Theory of Relativity.

Dear Merlin,

As I understand it the recession velocity of galaxies is proportional to their distance. I think this implies acceleration. What causes this acceleration? What is the rate of acceleration?

RON GIFFIN
CINCINNATI, OHIO

The word "acceleration," when used in the context of the expanding universe, is a misleading term. It is more precise to say "expansion rate."

The latest estimate for the expansion rate of the universe is about 15 miles per second for every million light-years from the Milky Way. This is the famous "Hubble constant" that was first estimated by Edwin Hubble in 1929.

The big bang is the source of the initial energy that set the universe expanding. Ever since then, the collective gravity of all the galaxies in the universe has actually *slowed* the expansion rate to its current value.

Dear Merlin,

If we look in one direction and see a galaxy receding at half the speed of light and we look in the opposite direction and see a galaxy receding at half the speed of light what would one galaxy say about the motion of the other? Would one observe the other to recede at the speed of light?

DAVID WILSON, M.D.

NEW ORLEANS, LOUISIANA

In standard pre-twentieth-century mechanics we would simply add the two speeds to get a relative velocity of the speed of light. We know from twentieth-century physics, however, that relative motion at speeds near the speed of light requires a bit more math.

If one of the galaxies knew all about Einstein's Special Theory of Relativity it would probably jot down the following on a cosmic notepad:

$$V_R = \frac{V_1 + V_2}{\left[1 + \frac{V_1 \times V_2}{c^2}\right]}$$

$$V_R = \frac{\frac{c}{2} + \frac{c}{2}}{\left[1 + \frac{\frac{c}{2} \times \frac{c}{2}}{c^2}\right]}$$

$$V_R = \frac{c}{\left[1 + \frac{(\frac{c}{2})^2}{c^2}\right]}$$

$$V_R = \cfrac{c}{1+(\frac{1}{4})}$$

$$V_R = \frac{4}{5}c$$

And if the galaxy could speak, it would say, "I see that other galaxy recede from me at four-fifths the speed of light."

XI

TIME, SPACE, and a SENSE OF WHERE YOU ARE

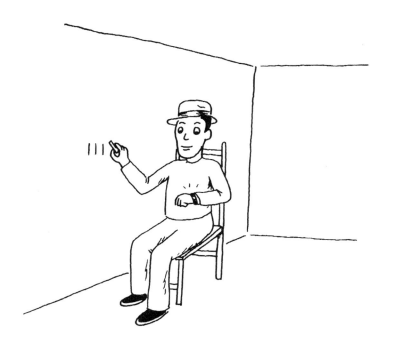

TIME, SPACE, AND A
SENSE OF WHERE YOU ARE

*L*ET'S MEET AT THE CORNER of 34th Street and Sixth Avenue on the eighty-sixth floor at 12:30 P.M. on the fifth of October in the year 2018.

This simple rendezvous we just arranged uses time-keeping and coordinate systems that uniquely identify a single place and time in the entire universe. But we must remember that calendars, clocks, numbered street grids, and even tall office buildings had to be invented.

Time and space coordinates together are needed for any rendezvous, but through the Theory of Relativity Einstein helped to fuse these two quantities into one conceptual entity called *space-time*. When combined with other principles, the Theory of Relativity unleashes a profound rethinking of the measurement of time, the fabric of space, the concept of mass, and the expansion of the universe.

Dear Merlin,

What is an eon? My Random House dictionary defines it as "the largest percent of geologic time comprising two or more eras." But my Webster defines eon as "an extremely long and indefinite period of time, thousands and thousands of years." I am confused.

EMI YAMAYISHI
PIEDMONT, CALIFORNIA

'Twas Merlin's Tour he wanted to be on.
Emi knew not how long was one eon.
Merlin said unto him,
"It's a billion years," with a grin,
"So don't you wait around to see one!"

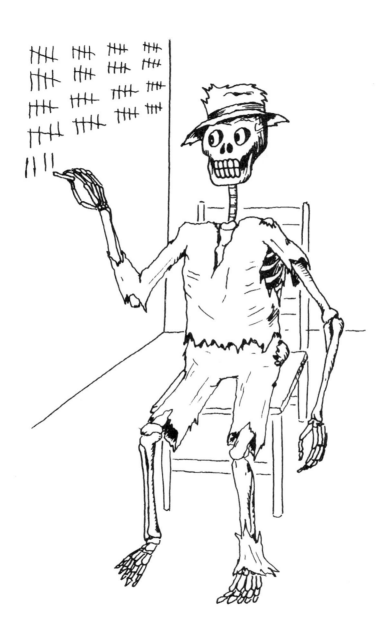

Dear Merlin,
  What is time?
    CARL LA VALLE
    PASADENA, CALIFORNIA

The period between two events or during which something exists,
happens, or acts.
WEBSTER'S NEW ENCYCLOPEDIC DICTIONARY OF THE
ENGLISH LANGUAGE

To everything there is a season
and a time to every purpose under the heaven.
ECCLESIASTES

If you can look into the seeds of time
And say which grain will grow and which will not,
Speak then to me, who neither beg nor fear
Your favours nor your hate.
WILLIAM SHAKESPEARE

Lives of great men all remind us we can make our lives sublime,
And, departing, leaving behind us footprints in the sands of time.
HENRY WADSWORTH LONGFELLOW

Time present and time past
Are both perhaps present in time future,
And time future contained in time past.
T. S. ELIOT

Time is defined to make motion look simple.
ALBERT EINSTEIN

Time is nature's way to prevent everything from happening all at once.
A BATHROOM WALL

Time is whatever your clock shows.
MERLIN

Dear Merlin,

When I was a boy things were regular. The first day of winter was December 21, that of spring March 21, etc.

Well, the bookkeepers seem to be all off nowadays. Seasons are beginning on the 20th, the 22nd, and even the 23rd. How come we can't keep the Sun crossing the equator on time anymore?

DAVID H. WALWORTH, M.D.
SAN JOSE, CALIFORNIA

$\mathcal{M}$erlin doesn't know what calendar you used when you were a boy but it certainly wasn't the same as that used by all of western society since the year 1582. (Unless, of course, you are 421 years old.)

Earth takes longer than 365 days to orbit the Sun (about 365¼ days) so the beginning of each season migrates ¼ a day per year. In February we add a leap day every four years to prevent the seasons from marching systematically through the entire calendar.

Dear Merlin,

It is said that we live in a four-dimensional universe. I have trouble picturing four dimensions in my mind but I want to understand the concept. Can you help me?

J. SHERMAN

CHATHAM, NEW JERSEY

$\mathcal{T}$here is nothing wrong with you just because you can't picture a four-dimensional universe. It is more important for you to understand that the fourth dimension, time, is given equal mathematical treatment to the three more familiar space dimensions.

Merlin thinks Hermann Minkowski said it best in 1908 when he commented, "Nobody has ever noticed a place except at a time or a time except at a place. . . . Henceforth space by itself, and time by itself, are doomed to fade away into mere shadows, and only a kind of union of the two will preserve an independent reality."

Dear Merlin,

If my twin were to visit your home on planet Omniscia in the Andromeda galaxy and return to Earth, my understanding is that he would age less rapidly than I. As my twin speeds away from me toward Omniscia and returns, so do I speed away from my twin and return. Therefore, don't I age less rapidly when viewed by my twin?

RON GIFFIN
CINCINNATI, OHIO

Your question refers to the famous "twin paradox" of Einstein's Theory of Relativity. Let's name your twin brother "Jon" and repeat the paradox in a little more detail.

Ron and Jon are identical twins on future Earth. Jon decides to visit Omniscia in the Andromeda galaxy two million light-years away. This is a long trip, but advances in space travel permit Jon to move at nearly the speed of light. According to Einstein's relativity, Jon will see Ron age less rapidly and Ron will see Jon age less rapidly. Motion at constant velocity is relative so Jon's and Ron's views are equally valid.

Jon doesn't stay long on planet Omniscia and returns to Earth to rejoin Ron. Along the way, once again, Jon sees Ron age less rapidly and Ron sees Jon age less rapidly. When Jon greets Ron back on Earth, who is older than who?

The paradox is resolved when we ask who actually made the round trip. Was it Jon and his spaceship or did the Earth and planet Omniscia slide through space two million light-years, and then slide back to where they started? The question sounds absurd, but the answer is not obvious.

In Einstein's relativity, motion at constant velocity is relative but accelerated (and decelerated) motion is not.

When Jon came to an abrupt stop on planet Omniscia his dishes rolled off the shelves in his space cabinet. No such phenomenon occurred on Earth because Earth didn't decelerate.

On Jon's return trip, he accelerated to near the speed of light and the ball he plays with rolled quickly to the rear of the space ship. At the same time, Ron didn't roll off the back of Earth because Earth didn't experience the acceleration.

In the twin paradox, Einstein's relativity shows that it is the accelerating and decelerating twin that ages the least.

Alas, Jon returns younger than Ron by several million years.

Dear Merlin,

If the key to time dilation is speed, then wouldn't we be included in it since Earth is traveling at 18 mi/sec around the Sun?

DOMINICK DESIMONE
STATEN ISLAND, NEW YORK

$Y_{es.}$

An observer on the Sun (equipped with a flameproof watch) will see Earth, its residents, and all time-keeping devices lose about one second in thirty years.

Dear Merlin,
   What is a tachyon?
      Rick McFairlane
      Dallas, Texas

A tachyon is a hypothetical particle that travels faster than the speed of light. Einstein's equations of special relativity bestow this particle with an array of bizarre properties:

1. The slowest a tachyon can move is slightly greater than the speed of light.
2. A tachyon can have an infinite velocity.
3. When a tachyon loses energy it speeds up—when it gains energy it slows down.
4. It would take infinite energy to slow down a tachyon to the speed of light.
5. Tachyons can penetrate easily through six miles of lead.
6. A tachyon appears to travel backwards in time for some observers. If you send your friends a message with tachyons they can receive the message *before* you send it.

Tachyons have yet to be detected.

Dear Merlin,

I understand that astronomers keep track of the exact time interval between dates by use of a "Julian Period," which begins January 1, 4713 B.C. Why was the year 4713 B.C. selected? Why not 4800 or 5000 B.C.?

ALVIN J. BUTOW
MARIPOSA, CALIFORNIA

The Julian Period is the number of days in 7,980 years beginning at 12 noon on January 1, 4713 B.C. The Italian Protestant scholar, Joseph Scaliger, introduced this time-keeping scheme in 1582—the first year of the Gregorian Calendar that is now the standard in most of the world's countries.

Scaliger named this period in honor of his father, Julius. It was designed to be independent of all other calendar schemes since it merely counted elapsed days.

There is nothing mysterious about the year 4713 B.C. It was chosen to be so long ago that no recorded astronomical event exists before it. Aside from this stipulation, the exact date was completely arbitrary. The modern day penchant for "round" numbers did not exist four hundred years ago.

The 7,980-year period was chosen as the smallest common multiple of the 28-year solar cycle, the 19-year Metonic lunar cycle, and the 15-year Roman cycle of indiction.

The Julian day for noon, January 1, 2000, is 2,451,545.000.

Dear Merlin,

When does the twenty-first century officially begin? Is it January 1, the year 2000, or is it January 1, the year 2001?

MARGARET M. SIPPLE

LOS ANGELES, CALIFORNIA

Calendar purists will tell you that the twenty-first century begins on January 1, in the year 2001. In the current Christian-based Gregorian calendar the year 1 B.C. was followed by the year A.D. 1—there was no year zero. Century reckoning is therefore shifted by one year.

Before you get excited about this please note a few things. 1) The "zero point" of any reckoning of time is arbitrary. 2) No day before October 10, 1582, actually occurred on the Gregorian calendar for that was when the calendar was implemented officially. And 3) the Gregorian calendar doesn't even start in the "right" place. Bible scholars generally place the birth of Christ several years *before* 1 B.C.

When we consider the arbitrary numerics of it all then it shouldn't really matter which January 1 you choose to celebrate the twenty-first century. Why not celebrate it twice?

Dear Merlin,
> *What is space? How is it defined?*
>> M A R K   V .   R A U B E R
>> H U M B L E ,   T E X A S

When NASA says they go into space they usually refer to sending a *space*craft in orbit around Earth or elsewhere in the solar system.

Merlin prefers to think of space as the regions between all the particles of all the atoms of the universe. With this definition matter can be viewed as the defining edge of space.

We are left with the rhetorical question, "Can space exist in the absence of matter, if matter defines the edge of space?"

Merlin leaves that one for you.

Dear Merlin,

How empty is empty space?

ARTHUR LEVY

HOUSTON, TEXAS

When a rabbit disappears into "thin air" at a magic show nobody tells you the thin air already contains over 10,000,000,000,000,000,000 (ten quintillion) atoms per cubic centimeter. The very best laboratory vacuum chambers have as few as 10,000 atoms per cubic centimeter. Interplanetary space gets down to about 10 atoms per cubic centimeter while interstellar space is as low as 0.5 atoms per cubic centimeter. The award for nothingness, however, must be given to intergalactic space. There it is difficult to find more than 0.0000001 atoms per cubic centimeter.

It has been postulated that outside the universe, where there is no space, there is no nothing. We might call this hypothetical region (where we are certain to find multitudes of rabbits) *nothing-nothing.*

Dear Merlin,
  How cold is space?
    BRUCE WALKER
    ARVADA, COLORADO

If you position yourself in intergalactic space to eat breakfast you will find that your pancakes, your coffee, the ice in your orange juice, and everything else around you will cool continuously until it reaches −454° Fahrenheit—the relic temperature of the universe from the big bang. On the Kelvin scale this temperature is more commonly called the "3° background radiation."

If you decide to journey close to a star, then the temperature your food reaches will be correspondingly higher.

Dear Merlin,

Is there any concern over the pollution of space? Space was a pristine wilderness until man, pursuing his scientific, military, and commercial endeavors, trashed it!

ED CARLSON
MARIPOSA, CALIFORNIA

Interplanetary space is not as pristine as you describe. Earth intercepts over one thousand tons of debris every day in the form of meteors and micrometeors. Fortunately, nearly all of it burns in Earth's atmosphere. (Let us not forget what the surface of the Moon looks like.)

Satellite and spacecraft debris also account for quite a bit of "pollution." We have formed a veritable traffic hazard for space travelers—especially since most of the orbiting debris is moving between ten and twenty thousand miles per hour.

To keep travel safe for the future, Merlin believes that all jettisoned debris should be broken into little pieces and then sent to reenter and burn through the Earth's atmosphere. Interplanetary travelers could jettison their debris toward the Sun.

By this scheme, all spacecraft would "dispose" of their own garbage.

Dear Merlin,

Is it true that what goes up must come down?

MARIANNE FARMER
OMAHA, NEBRASKA

*N*o.

Some of what goes up,
If launched with great ferocity,
Won't ever give up—
It reached escape velocity.

Some of what goes up,
If launched high in the sky,
Won't ever return—
Air friction makes it burn.

The rest of what goes up,
Launched slowly from the ground,
Started the old adage,
"What goes up, must come down!"

Dear Merlin,

Are there directions such as east and west or up and down in space?

ANNA DENHAM
YUKON, OKLAHOMA

It is important for Anna to test
That one man's east is another man's west.
And in space, since there is no up or down,
One man's smile will be another man's frown.

Dear Merlin,

Can you name some molecules that we know to exist in space? I hear that elements form complex molecules quite often in interstellar clouds.

JANA BAILEY
HARRISBURG, PENNSYLVANIA

Cold interstellar clouds are rich with heavy elements that readily form complex molecules.

Some are familiar "household" molecules:

NH_3 (ammonia)
H_2O (water)
C_2H_2 (acetylene)

And some are deadly:

CN   (cyanogen)
CO   (carbon monoxide)
HCN   (hydrogen cyanide)

Some remind you of the hospital:

H_2CO (formaldehyde)
C_2H_5OH (ethyl alcohol)

And some don't remind you of anything:

N_2H^+ (dinitrogen monohydride ion)
HC_5N (cyanodiacetylene)

At last count more than fifty types of molecules have been discovered.

Dear Merlin,

What does it mean in Einstein's General Theory of Relativity for space to curve? How can space curve?

WALTER SCHATZ, M.D.

SAN DIEGO, CALIFORNIA

According to Isaac Newton, gravity is a force that acts at a distance between any two masses.

According to Albert Einstein, however, gravity is the curvature of space. Your speed in the vicinity of a mass will determine how much you respond to its curved space. A planetary orbit, for example, is described quite simply as the response of a planet to the curvature of space in the vicinity of the Sun. The speed of light is large enough so that its path is affected only minimally in the presence of the Sun.

But space becomes severely curved in the vicinity of exotic gravitational objects like black holes and neutron stars. In a carefully chosen trajectory you can send light into orbit around a black hole.

On a much larger scale, the entire space of the universe is curved in response to the collective gravity of nearly a hundred billion galaxies. Merlin overheard Einstein comment one day, "Matter tells space how to curve, and space tells matter how to move."

Dear Merlin,
Over the Earth's 26,000-year precession, aren't the current values for right ascension constantly (albeit slowly) changing? It seems odd that a reference value is variable.

JIM RICHARDS
LIVERMORE, CALIFORNIA

The celestial coordinate system that astronomers use (right ascension and declination) is indeed changing continuously. At the large telescopes that professional astronomers use there are "on-board" computer programs whose duty it is to convert coordinates.

You feed the computer the coordinates of an object and the date for which the coordinates are valid. These are taken routinely from standard reference tables or star charts. The program then "precesses" the old coordinates to the new coordinates appropriate for the moment of your observation.

This change is normally small and would go unnoticed to the unaided eye during a human life span.

Dear Merlin,

I was reading an article about IRAS 1349 + 2438. I wondered just what these numbers meant or how they were selected; especially the "plus" and additional numbers.

A<small>LICE</small> K. P<small>ARKER</small>
L<small>EWES</small>, D<small>ELAWARE</small>

*T*he universe brims with mystery.

Most catalogues of astronomical objects list entries in that format.

If we were to make a sentence out of it, the sentence would read, "Infrared Astronomical Satellite catalogue— Object detected at 13 hours 49 minutes Right Ascension and +24° 38′ Declination on the celestial sphere." These coordinates are the astronomers' counterpart sky grid to the Earth-based longitude and latitude system.

Dear Merlin,

We know that Earth is located in the solar system, and the solar system is located in the Milky Way galaxy, and the Milky Way galaxy is located in the universe. But where is the universe located?

JIM SCHLIGHT
WEST JORDAN, UTAH

To answer your question is to leave the domain of scientific inquiry and enter the realm of metaphysics.

Merlin is happy to give you one example:

The universe is the contents of a fifteen-billion-light-year-diameter black hole that is embedded in a parallel universe of which we are a microcosmic subset.

Notice the above statement has little basis in scientific data and it cannot be verified (since by definition we cannot communicate with regions outside the universe). It is storytelling, not science.

XII

Black Holes, Quasars, and the Universe

BLACK HOLES, QUASARS, AND THE UNIVERSE

WHAT IS NOW PROVED
WAS ONCE ONLY IMAGIN'D.

—WILLIAM BLAKE

Dear Merlin,
   Exactly what is a black hole? How big are they? What are they made of?

FRANKIE CINTRON
BAYAMON, PUERTO RICO

A black hole can be any size.

What all black holes have in common is that light cannot escape them. Their gravity is so strong that, among other things, space curves back on itself so that even light cannot escape. If light cannot escape, then nothing can escape.

Some black holes are the final evolutionary stage of a star with more than about ten times the mass of the Sun.

Quasars and many "active" galaxies are likely to have supermassive black holes (up to a million times the Sun's mass) in their center. They devour stars that wander too close.

Higher mass black holes are bigger than lower mass black holes. As black holes eat they grow.

A journey to a black hole is a one-way trip. You may wish to avoid them.

Dear Merlin,

I don't understand how a black hole could become so dense that it could be the size of an atom.

ERIN FRENCH

LANSING, MICHIGAN

*N*either does anybody else.

Einstein's equations of relativity quite naturally account for the never-ending collapse of a black hole—in spite of how much common sense is defied.

If there exists a force to prevent the collapse of a black hole then it is yet to be discovered.

Dear Merlin,

How does a black hole affect time and mass?

TERRY ANSTEY
AUSTIN, TEXAS

*E*instein's General Theory of Relativity describes the curvature of space-time in the vicinity of a black hole. The equations show that if you fell toward a black hole your time slows down and your mass increases as viewed by an observer perched a safe distance away. As you near the edge of the black hole, the "event horizon," your time almost stops, and your mass increases without limit.

Dear Merlin,
    What would happen to me if I fell into a black hole?
        BENNY JACKSON
        TULSA, OKLAHOMA

In your feet-first dive
To this cosmic abyss,
You will not survive
Because you will not miss.

The tidal forces of gravity
Will create quite a calamity
When you are stretched head-to-toe.
Are you sure you want to go?

Your body's atoms—you will see them,
Will enter one-by-one.
The event horizon will eat 'em.
You won't be having fun.

Dear Merlin,

If the Sun became a black hole, what would happen to Earth's orbit?

R. C. RITTER
EL PASO, TEXAS

The familiar properties of Earth's orbit are only related to the total mass of the Sun. If the Sun were compacted to form a black hole then its mass would remain unchanged and Earth's orbit would be unaffected and sunrise would be uninteresting.

Dear Merlin,

I understand how pulsars got their name. They are stars that emit light pulses. How did quasars get their name? I presume the astronomical object preceded the Quasar brand television.

HILLARY SMEDMAN

DAYTON, OHIO

In 1963 an object was discovered in the Third Cambridge Catalogue of radio-emitting objects that looked like a star on a photograph but emitted radio waves of enormous intensity. Its redshift implied it was farther away and more luminous than any known galaxy—it could not be a star. It was aptly described as a quasistellar radio source, or "quasar" for short.

The Quasar brand television appeared somewhat later. Incidentally, astronomy can also be credited with the following product names: *Pulsar* watch, *Galaxy* fan, *Mars* and *Milky Way* candy bars, *Comet* cleanser, *Celestial* Seasonings tea, *Moonglo* bath oils, *Eclipse* paint, *Universal* Studios, and Chevy *Nova*. (Although if Chevrolet had known that a nova is a star that has just blown up they might have reconsidered the name of their vehicle).

Dear Merlin,

Quasars are noted for their great distance and luminosity. Are there any quasars in our galaxy? How far away is the nearest one?

DUANE MORSE

PHOENIX, ARIZONA

*I*f there were any quasars in our own galaxy we would certainly know about it (by way of death from high-energy radiation). Quasars are found to have hundreds and sometimes thousands of times the luminosity of the entire Milky Way.

The nearest full-scale quasar is 3C273, a safe $1\frac{1}{2}$ billion light-years away.

Nearer than this, however, are galaxies with "active" nuclei that share some of the properties of the classical quasars but do not quite have the enormous radio or optical light output.

Dear Merlin,

If our astronomers can see a quasar ten billion light-years away that recedes at 70 percent of the speed of light, may I conclude:

1. that the quasar, today, has receded for another ten billion years;
2. that its speed and redshift are so great that we will never see it ten billion years from now;
3. that if the universe contracts, we should begin to see some quasars approaching;
4. that I should have my head examined.

FRANK LOESCH
PITTSBORO, NORTH CAROLINA

*M*erlin concludes:

1. that yes, the quasar today has receded for another ten billion years;
2. that yes, in an infinite universe, some quasars today will not be detected because they have reached the event horizon of the universe where their spectra are infinitely redshifted;
3. that yes, if the universe contracts, quasars (and the rest of the contents of the universe) will approach us and display a blue shift;
4. that the only people who should have their head examined are those who never ask questions.

Dear Merlin,

What is the notion of missing mass in the universe? How do you know it was supposed to be there to begin with and how do astronomers know it is missing?

LUKE WEBBERMAN
STATEN ISLAND, NEW YORK

The "missing mass" problem, also known as the "dark matter" problem, was first identified by the astronomer Fritz Zwicky in the 1930s. It arises most prominently in the case of galaxy clusters.

To determine the mass of a cluster the most straightforward way is to add up all the light from the individual galaxies. If the relation between the light detected and the underlying mass is known then you can infer a total mass for the cluster.

There is another way to estimate the mass of a galaxy cluster. It assumes that the orbital speeds of the galaxies in the cluster (a readily determined value) are related to the cluster's total mass—a direct consequence of Isaac Newton's Law of Gravity. This method yields mass estimates up to one hundred times *larger* than the first method. The theoreticians call it the "missing mass" problem while the observational astronomers call it the "missing light" problem.

The discrepancy is still unresolved.

Dear Merlin,

Where is the theoretical center of the universe in the big bang theory?

TASKER N. RODMAN, M.D.
LEACHVILLE, ARKANSAS

*W*ith the advent of relativity in 1905, Albert Einstein and Hermann Minkowski showed mathematically that time is as much a dimension as space. To ask "where?" is also to ask "when?" To help digest the proper answer to your question we should first think in three dimensions—two space dimensions and one time dimension.

Imagine galaxies etched on the two-dimensional surface of a balloon. As the balloon expands, Merlin can ask you, "Where is the center of the balloon?" You will probably say in the middle—inside the balloon. But the balloon's surface, where we displayed our etchings, was at the center only when you first started to inflate the balloon. The center of the balloon does not exist on its surface, except back in time at the balloon's origin.

By comparison, in our four-dimensional universe—three space dimensions and one time dimension—you ask, "Where is the center?" Merlin can legitimately reply, "It happened everywhere in space at $T = 0$, the beginning of time and the beginning of the universe."

Dear Merlin,

What is the radius of the known universe? Is it theoretically limited?

SPENCER S.

SANTA BARBARA, CALIFORNIA

In our expanding universe, the far galaxies are receding faster than the near ones. This is one of the signatures of the big bang explosion.

The radius of the observable universe may then be defined as the distance a galaxy would have to be for it to recede from the Milky Way at the speed of light. This distance is the "event horizon" of the universe.

The exact value is still debated, but all the latest estimates fall between twelve and sixteen billion light-years.

Dear Merlin,

What was around before the big bang brought the universe into existence?

MARTHA SPRINGER
DALLAS, TEXAS

Time is defined to have begun at the instant of the big bang. Merlin may be old, but Merlin is not older than time itself.

Observational astronomy and all known laws of physics describe the appearance and behavior of the universe after the big bang. There does not exist an observation, an experiment, or a morsel of data that would lend a clue to answer your question without entering the realm of metaphysics.

Dear Merlin,

If all mass is the result of the big bang fifteen billion years ago (all mass at the same location before the expansion), then light emitted from a galaxy fifteen billion years ago was emitted "here," not fifteen billion light-years away where the galaxy is now. How can we see light from fifteen billion years away that was generated fifteen billion years ago if the galaxy was, in effect, "here" then?

STEPHEN C. BROWN
INVERNESS, FLORIDA

You are confusing your heres and theres with your nows and thens.

When we see the light of the galaxies at the far reaches of the universe we see the universe not as it *is* nor where it *is*. We see it as it *was* and where it *was*.

The only way there can be a cosmic "now" is if everybody is in the same place. This occurred in the first moment of the big bang and has never happened since then.

Remember that observers on remote galaxies see us (the Milky Way galaxy) near the edge of the universe as well. But the light they see "now" was what we looked like ten billion years ago.

There is no doubt that someone on one of these remote galaxies is asking Merlin's cousin the same question you are.

Dear Merlin,

If the universe is infinitely large then isn't the light from another one or two hundred billion galaxies speeding on its way toward Earth right now? If that is so then won't moonless nights soon be as bright as moonlit nights?

GEOFFREY A. GODFREY

VENTURA, CALIFORNIA

If the universe were infinitely large and infinitely old and not expanding then the nighttime sky would be ablaze with light from the collective luminosity of all galaxies. But the night sky is dark. (Unless, of course, you live near downtown Los Angeles.) This age-old problem was first formulated by Heinrich Wilhelm Matthäus Olbers in 1826. It is named "Olbers's paradox" in his honor.

We may resolve the paradox by noting two important features of the cosmos. About fifteen billion light-years away lies the visible edge of the universe—it is not infinitely large. Also, the universe is expanding, which has a diluting effect on the intensity of the light en route to Earth from the distant galaxies. Both these effects permit all residents of the universe to have dark, romantic nights.

Dear Merlin,

You tell us that space around us is expanding more slowly than space at the outer edges of the universe, so I presume that means that the wave front of the big bang is at the outer edges of our space. Now if the residual radiation as detected from Earth is about 3° Kelvin, and if my presumption is correct, to a certain extent would the residual radiation near the wave front be higher than our local measurements?

FRANK W. STONES
FORT WORTH, TEXAS

No.

The residual background temperature of the entire universe is *now* about 3° Kelvin. Since you live in the present, this is what you measure.

When you observe the rapidly expanding outer reaches of the universe you see the universe as it used to be. If you were alive in that epoch you would measure a considerably hotter universe since the universe was smaller and the big bang would have occurred more recently. Only in that sense could you be considered closer to the "wave front."

Dear Merlin,

If my spaceship approaches the speed of light, could I escape the expansion of the universe? Does it depend on whether the universe is infinite or finite?

MIKE HAMILL
AUSTIN, TEXAS

$\mathcal{M}$erlin doesn't know why you want to escape the universe.

If the universe is infinite it will expand forever and you will never catch up with that part of the universe that is already receding from you at the speed of light.

If the universe is finite, it will one day recollapse and your journey to the "edge" of the universe will return you to where you started for you will have traveled in a gigantic cosmic circle.

If you must leave the universe, then the easy way to do it is to fall into a black hole. You don't even need a spaceship—the gravity will pull you straight in. Communication with the outside cannot occur from within the event horizon and you will never come out. You will have left the universe forever.

Dear Merlin,

As our universe expands, the total amount of "useful" energy continually decreases, passing into dissipated, or thermal energy (entropy increases). If there is enough "missing mass" to cause the universe to one day begin contracting into the "big crunch," does this mean that all physical processes will then lead to an increase in useful energy (a loss of entropy)?

GEOFFREY G. HUGGINS
WINCHESTER, VIRGINIA

Nobody knows for sure, but there are two schools of thought if the universe recollapses:

One says that all laws of physics will continue as they are (including the law of increasing entropy) during the expansion and contraction of the universe.

The other school holds that the lost energy (increased entropy) in the expansion will be regained in the contraction.

Merlin prefers the second school.

Dear Merlin,

What is the farthest quasar in the universe that astronomers have seen to date?

RAYMOND TOLAR
MONTREAL, QUEBEC

The last Merlin checked, the farthest known quasar in the universe was PC1247 + 3406 with a redshift of 4.897. It is found in the constellation Canes Venatici, the Hunting Dogs.

In the general expansion of the universe it is receding from the Milky Way galaxy at nearly 95 percent of the speed of light and is at an estimated distance of over twelve billion light-years.

Dear Merlin,

If the Moon, Earth, planets, stars, and galaxies rotate, does the universe rotate?

BRAD WILKINSON

SAN FRANCISCO, CALIFORNIA

There exists no evidence to support this idea.

If the universe were rotating it would go unnoticed only if our Milky Way galaxy were at the center of all rotation. Redshift surveys of galaxies in the universe only detect motion toward or away from us. The component of a galaxy's motion that is sideways to our view would be unmeasurable by all existing observational methods.

If the Milky Way were positioned anywhere other than the center of a rotating universe then the redshift surveys would show large sections of the universe approaching us, other sections receding from us, and still other sections that are neither approaching nor receding.

This is not detected.

XIII

LIFE:

HERE

and

THERE

LIFE: HERE AND THERE

THE MOST FREQUENTLY ASKED question in all of astronomy appears in this chapter. There seems to be a basic and worldwide need to know whether life exists on other planets. To appreciate the barely fathomable size and contents of the universe is to recognize that it is imprudent to believe Earth is the only planet with life.

It is not clear what extraterrestrial life forms would be like. There is still very much about life on Earth that remains to be determined. Regardless of this ignorance, life on Earth remains the one example of life as we know it (LAWKI). Any experimental search for life is likely to have this inherent bias. Even in science fiction films it is common to show grotesque (or cute) aliens that still have two eyes, a nose, a mouth, a head, a neck, arms, and fingers. No matter how ugly the alien is, it often resembles a human more than a human resembles other forms of life on Earth like a jellyfish, an amoeba, or a Venus flytrap.

The next millennium of Earth civilization may herald interstellar space travel, thus increasing the chance of contacting extraterrestrial life other than Merlin.

Dear Merlin,

Has the growth of plants and animals had any appreciable effect on the gross weight of the Earth?

J. O. Edison
Brooklyn, New York

The "gross weight" of Earth, if you include the oceans and atmosphere, remains unaffected by the growth of plants and animals.

The oceans supply the atmosphere with water. The atmosphere supplies the ground with rain. The plants absorb the rain from the ground and grow. Some animals grow from the plants they eat. Other animals grow from the animals they eat. In this closed ecosystem all weight is accounted for.

Dear Merlin,

What would happen if there were no Moon? How would this affect Earth, the oceans, the weather, plant life, and the things on Earth?

JOSEPH KOLGAT, JR.
MILWAUKEE, WISCONSIN

If the Moon were to go away we don't expect the weather or climate to be much affected, nor plant life, nor geological processes (like mountain building, earthquakes, etc.).

The tidal sloshing of the oceans against the continental shelves, however, would be reduced considerably. (The smaller tidal effects from the Sun still remain.) In turn, the slowing of Earth's rotation rate would be reduced.

The rising and falling of the tides is historically important for the transition of sea life, through its amphibious phase, to land animals. Without the Moon, Merlin may well have been answering questions posed by fish.

Other miscellaneous effects include: The books and lore on werewolves would have never been written. The time interval "month" would either not exist or have another name. There would have been no NASA Apollo missions. And "lunatics" would be in the market for a new identity.

Dear Merlin,
    How would the Sun affect Earth if Earth lost its ozone layer?

RICHARD BELOTE
WIMBERLEY, TEXAS

Ozone is a simple gas molecule that contains three oxygen atoms. It is found primarily in the upper atmosphere, where it absorbs over 99 percent of the ultraviolet (UV) and X-rays from the Sun. If, somehow, Earth lost its protective ozone then a chain of events would unravel that is likely to redetermine the ecological balance of life as we know it.

- Indoor tanning salons would go out of business immediately.
- Humans, especially light-skinned people, would become "sunburned," contract skin cancer, and risk irreversible retina damage.
- Crop yield would be reduced significantly. The germination and insect resistance in many plants is adversely affected by exposure to UV light. The list includes soybeans, corn, wheat, and rice. These four foods represent over 65 percent of the total caloric intake of the world's seven billion humans.
- Phytoplankton would die completely, thereby knocking out the base of the oceanic food chain as well as removing an important source of atmospheric oxygen.

A new ecosystem would emerge that is dominated by the remaining UV-resistant life forms. But this would not occur before most of the human population and much of the fish population starved to death.

In the time it took you to read this answer over a thousand pounds of ozone-destroying chlorofluorocarbons were released into the atmosphere by underarm deodorant sprays, paint sprays, escaped refrigerants, and hair sprays.

Dear Merlin,

Do you know of any other star, besides our Sun, that has a solar system with planets?

JAMES CONTRADA
MALVERNE, NEW YORK

Most stars in the universe are expected to have planets. But planets emit no visible light of their own— just reflected light from the star they orbit.

External solar systems that are just forming, however, emit (invisible) infrared light that is detectable from orbit by specially designed infrared telescopes. Just a few stars have been discovered to have this kind of debris (clouds, particles, etc.) in orbit around them. The best known among them is the bright star Vega in the constellation Lyra.

Other planetary systems have been discovered by monitoring the motion through space of the host star. If it jiggles, another object's gravity is responsible. About a dozen of these systems are now known, and the number is rising.

Dear Merlin,
    What is the statistical probability of other planets similar to
Earth existing in other solar systems, and even other galaxies?
        STEPHEN BROWNING
        FORT WORTH, TEXAS

Since the Sun (an unremarkable star) has nine planets,
we consider planets to be quite common in the universe.
Lets assume that every star has ten planets to make the
math simple.

    There are approximately
1,000,000,000,000,000,000,000 (one sextillion) stars in all
the galaxies of the universe. About half of all these stars
are part of double or multiple systems that render a
planet's orbit unstable. About half of the remaining stars
were formed in the early universe when no elements
heavier than hydrogen and helium were available to form
planets.

    A special feature about the planet Earth (and
Merlin's home planet Omniscia) is that it is in the right
distance-zone from the Sun to sustain *liquid* water, an
essential ingredient to life as we know it. The "width"
of this special zone is extremely narrow for the stars that
are much cooler than the Sun. They are poor candidates
for Earthlike planets. Ninety percent of all stars fit this
category.

    For the rest of the stars, if only about one planet in
ten (using your solar system as a model) can sustain

liquid water then Merlin's abacus shows we are left with about 25,000,000,000,000,000,000 (25 quintillion) Earthlike planets in the universe.

Estimating the fraction of these planets with "intelligent" life is considerably more speculative.

Dear Merlin,
    Could the rings around the other planets of the universe hold remnants of man-made objects from a civilization that inhabited one of the exploded planets?
        C. W. BARNES
        DALLAS, TEXAS

Merlin has never seen a planet explode, nor does Merlin know of any physical reason why a planet should explode, nor has Merlin ever seen "man-made" objects in another civilization.

    But if there were aliens on another planet that owned "alien-made" objects, and if their planet exploded, and if the explosion didn't vaporize their cultural artifacts, and if the debris came in the vicinity of another planet, and if the debris had just the right speed and angle of travel to be captured in orbit around this other planet, then yes, it is possible to find archaeologically interesting objects in the rings of some planets.

Dear Merlin,

I like to read about the possibility of planets elsewhere in the universe that are similar to Earth. Given the unwillingness and reluctance of people in this and other countries to accept persons of different race or ethnic background even though we are basically the same, how do you think we would receive intelligent life from another world that might be totally different from anything we know?

WILLIAM JAMES
CINCINNATI, OHIO

T here is some hope.

Earthlings seem to have accepted Merlin from the Andromeda galaxy. They even write to Merlin at:

Merlin
c/o *Star Date*
McDonald Observatory
The University of Texas at Austin
Austin, Texas 78712

when they have questions about astronomy and space.

POSTSCRIPT

If you are now a little more
  cosmic'ly enlightened,
And if you are now a bit more
  scientifically tuned,
And if you now have a fellowship
  with the universe,
Then it will be no effort for you to
Keep looking up.

Merlin

APPENDIX

Number Names

1	one	10^0
10	ten	10^1
100	one hundred	10^2
1,000	one thousand	10^3
1,000,000	one million	10^6
1,000,000,000	one billion	10^9
1,000,000,000,000	one trillion	10^{12}
1,000,000,000,000,000	one quadrillion	10^{15}
1,000,000,000,000,000,000	one quintillion	10^{18}
1,000,000,000,000,000,000,000	one sextillion	10^{21}
	.	
	.	
	.	
1 followed by 100 zeroes	googol	10^{100}
1 followed by a googol zeroes	googolplex	10^{googol}

GLOSSARY OF SELECTED TERMS

ATOM: The smallest part of chemical element that retains the identity of the element. It is normally composed of electrons, protons, and neutrons.

AURORA AUSTRALIS (SOUTHERN LIGHTS): Produced when charged particles that are emitted by the Sun spiral in toward Earth's south magnetic pole and collide with molecules in Earth's upper atmosphere. These collisions temporarily leave the air aglow with dancing curtains of light.

AURORA BOREALIS (NORTHERN LIGHTS): The same as aurora australis; however, the curtain of light is produced by charged particles that spiral in to Earth's north magnetic pole.

BIG BANG: The scientific description of the origin of the universe. Its basic premise is that the universe began in an explosion that brought space and matter into existence approximately ten billion years ago. The universe, today, is still expanding from this explosion.

BLUE SHIFT: The shortening of the measured wavelength of light due to the motion of the light-emitting object toward you. Since motion is relative, this shift also occurs if you are in motion toward the object.

CELESTIAL EQUATOR: The projection of Earth's equator on the *celestial sphere*.

CELESTIAL SPHERE: The entire sky as though its contents (stars, planets, Sun, Moon, etc.) were embedded in the inside of a giant sphere with Earth in the center. It is an extremely useful concept for identifying the coordinates and positions of celestial objects.

CELSIUS TEMPERATURE SCALE: Named for Anders Celsius (1701–1744), the chemist who, in 1742, first developed the temperature scale where water freezes at 0° and boils at 100°. Also called "centigrade."

CENTER OF MASS: In most cases it may also be considered the center of gravity. It is the position inside an object or among several objects where the total weight balances. The system could conceivably be supported in *equilibrium* at this single point.

CENTIGRADE TEMPERATURE SCALE: The previous name for the Celsius scale. Used until 1948 when it was changed to "Celsius" by decree of the ninth General Conference on Weights and Measures. See also *Celsius temperature scale.*

CENTRIFUGAL FORCE: The outward force that an object feels when it revolves around any other object or position. It may also be considered the force that creates the tendency to "fly" off at a tangent. Strictly speaking, centrifugal force is not a true force at all. It is just the tendency for the revolving object to move in a straight line—which is what it would do if no force were acting to keep it revolving such as a tie-line or gravity.

CENTRIPETAL FORCE: The force that keeps an object revolving around any other object or position. For example, the Sun's gravity provides the centripetal force that keeps all planets in orbit. Otherwise they would fly away into interstellar space.

CONSTELLATION: The random patterns of stars in space as seen from the Earth. The *celestial sphere* has been segmented completely to denote the presence of eighty-eight constellations. Each constellation has a name that, in rare cases, actually describes the star pattern.

CORONA: The thin and vacuous outer atmosphere of the Sun, which has a temperature that is estimated to be millions of degrees. It is considerably dimmer than the visual surface of the Sun so its presence can only be detected with specially designed telescopes called "coronagraphs," or during a total solar eclipse.

COSMIC RAY: Fast moving charged particles that have enormous energy. They permeate all space and are capable of inducing genetic mutations. Their origin is still unknown.

CUBIC INCH, CUBIC CENTIMETER, CUBIC FOOT, ETC.: These are units of volume that can be remembered because a cube has volume—so that a "cubic" anything is the volume measured using your favorite unit to multiply the length, width, and height.

DECLINATION: The counterpart of Earth's *latitude* lines on to the *celestial sphere*. The Earth's *latitude* is given in degrees north or south of the equator whereas declination is given in positive and negative degrees.

DOPPLER SHIFT: Named for Christian Doppler (1803–1853), whose work in the mid-nineteenth century pioneered the study of the measured change in the pitch (frequency) of a sound as the sound-emitting object approaches or recedes. This shift in frequency was later determined to be a general phenomenon that occurs with any form of wave.

DUST CLOUD: *Gas clouds* in interstellar space that are cool enough for atoms to combine to form larger molecules. When this happens with carbon atoms the gas cloud is often called a dust cloud.

DYNAMICS: The study of the motion and the effect of forces on the interaction of objects. When applied to the motion of objects in the solar system and the universe it is often called celestial mechanics.

ELECTROLYSIS: The breaking of atomic bonds by electrical current.

ELECTRON: The subatomic particle that is negatively charged. It is found in equal numbers with the positively charged protons throughout the universe.

ELEMENTS: The basic component of all matter. All matter in the universe is composed of ninety-two elements that range from the smallest atom, hydrogen (with one proton in its nucleus), to the largest naturally occurring element, uranium (with ninety-two protons in its nucleus). Elements larger than uranium are produced in laboratories.

ENTROPY: A measure of the disorder in a system. For closed systems, where there is no outside influence, entropy has been experimentally determined always to increase. Wherever entropy decreases (i.e., order and structure increases), it is always at the expense of some external system that supplies energy.

EQUILIBRIUM: A chemical or dynamical state where the measured values of all relevant quantities remain the same over time.

EQUINOX: The two days in the year where the center of the Sun's disk crosses the celestial equator—the projection of the Earth's equator on the *celestial sphere.* The first of these two days occurs in March and is called the vernal equinox, or the first day of spring. The second occurs in September and is called the autumnal equinox, or the first day of fall.

ESCAPE VELOCITY: There is a special speed on all planets, stars, or anything with gravity where a tossed object will never return. This speed is defined as the escape velocity. For all speeds less than escape velocity the tossed object will return.

EVENT HORIZON: The poetic name given to the bounding region around a black hole within which light cannot escape. It may be defined as the "edge" of a black hole. This term is also applied to the visible edge of the universe.

FAHRENHEIT TEMPERATURE SCALE: Named for Gabriel Fahrenheit who first described the scale where water freezes at 32° and boils at 212°.

FISSION: The splitting of larger atoms into two or more smaller atoms. If this occurs with atoms larger than iron then energy is released. This is the source of energy in all present-day nuclear power plants. Also called "atomic fission."

FUSION: The combining of smaller atoms to form larger atoms. If this occurs with atoms smaller than iron then energy is released. The primary energy source for the world's nuclear war arsenals and for all stars in the universe is fusion. Also called *"thermonuclear* fusion."

GALACTIC CLUSTERS: Star clusters of all ages that were born in the disk of a spiral galaxy. They tend to contain less than several thousand stars. (Compare with *globular clusters.*) The presence of large *gas clouds* in the disk of most spiral galaxies ensures that galactic clusters will continue to be born.

GAS CLOUD: Clouds of hydrogen, helium, and trace amounts of heavier elements. These are the primary component of interstellar space in the disk of spiral galaxies.

GENERAL RELATIVITY: Introduced in 1915 by Albert Einstein, and published in 1916, it forms the natural extension of *Special Relativity* into the domain of accelerating objects. It is a modern theory of gravity that successfully explains many experimental results that were not otherwise explainable in terms of Newton's theory of gravity from the seventeenth century. Its basic premise is the "equivalence principle" whereby a person in a space ship, for example, cannot distinguish whether the space ship is accelerating through space, or whether the space ship is stationary in a gravitational field that would produce the same acceleration. From this simple yet profound principle emerges a completely reworked understanding of the nature of gravity. According to Einstein, gravity is not a force in the traditional meaning of the word. Gravity is the curvature of space in the vicinity of a mass. The motion of a nearby object is completely determined by its velocity and the amount of curvature that is present. As counterintuitive as this sounds, it explains all known behavior of gravitational systems ever studied and it predicts a myriad of even more counterintuitive phenomena that are continually verified by controlled experiments. For example, Einstein predicted that a strong gravity field should warp space and noticeably bend light in its vicinity. It was later shown that starlight passing near the edge of the Sun (as seen during a total solar eclipse) is found to be displaced from its expected position by an amount in exact accord with Einstein's predictions. Perhaps the most grand application of the General Theory of Relativity involves the description of our expanding universe, where all of space is curved from the collected gravity of hundreds of billions of galaxies. An important and currently unverified prediction is the existence of "gravitons" or "gravity waves." These are the particles of gravity that communicate abrupt changes in a gravitational field such as is expected in a supernova explosion.

GIBBOUS: The phase of the Moon (or any round object that goes through phases) that is found between quarter phase and full phase. If the phase is growing to be full then it is waxing gibbous. If the phase is diminishing toward last quarter it is called waning gibbous.

GLOBULAR CLUSTER: An old, large, gravitationally bound collection of stars that is roughly spherical in shape. In spiral galaxies, like the Milky Way, globular clusters are found principally in the *halo*. They are also found in elliptical galaxies.

HALO: The large spherical region around the Milky Way galaxy that contains the *globular clusters* of stars. It is mostly devoid of interstellar gas and dust.

INDICTION, CYCLE OF: The fifteen-year cycle for figuring property value to be taxed during the late Roman Empire. It is the only cycle in the Julian Period that has no astronomical basis.

INTERNATIONAL ASTRONOMICAL UNION (IAU): The largest governing body for professional astronomers in the world. Among other things, they sponsor conferences and workshops on special research topics.

KELVIN TEMPERATURE SCALE: Named for Lord Kelvin (1824–1907). He invented the scale where the coldest possible temperature is, by definition, 0°. Its increments are the same as the Celsius scale. On the Kelvin scale water freezes at 273.16° and boils at 373.16°.

KINETIC ENERGY: The energy that an object has by virtue of its motion. For objects in motion, mass also contributes to kinetic energy. For example, if a more massive object (like a truck) moves with the same speed as a less massive object (like a tricycle) then the more massive object will have more kinetic energy.

LATITUDE: On Earth it is the coordinate that measures how far you are in degrees from the equator. The equator is defined as zero degrees and the poles are 90 degrees north and 90 degrees south.

LOCAL GROUP: The friendly name given to the twenty or so galaxies in the immediate vicinity of the Milky Way galaxy. The Local Group includes the Magellanic Clouds and the Andromeda Galaxy.

LOGARITHMIC SCALE: A method for plotting data whereby tremendous ranges of numbers can fit on the same piece of paper. In official terms, the logarithmic scale increases exponentially (e.g., 1, 10, 100, 1,000, 10,000) rather than arithmetically (e.g., 1, 2, 3, 4, 5).

LONGITUDE: On Earth, it is the coordinate that measures how far east or west you are from the arbitrarily defined "prime meridian" that goes north–south through Greenwich, England. You can venture up to 180 degrees east or 180 degrees west of Greenwich, thus spanning the 360 degrees around the Earth.

LUMINOSITY CLASS: A classification scheme for the size and overall luminosity of a star. Class I refers to super giants, Class III refers to ordinary red giants, and Class V refers to stars that have not yet reached the giant phase and are still converting hydrogen to helium in their core.

MAGNETIC FIELD: Moving charged particles are the sole producers of magnetic fields. These fields are regions of space that supply a force to other charged particles in the area. All magnets have two poles that are often called north and south. If you represent a magnetic field with imaginary lines then all lines form complete loops that extend through both magnetic poles.

MASS: A measure of an object's material content. For example, a locomotive that is weightless in space has no less mass than a locomotive that weighs one hundred tons on Earth. Note also that mass makes no reference to size. A beach ball is large but it is certainly not massive. An anvil is massive, but it is certainly not large.

METABOLISM: The rate that a living creature uses energy. A high metabolism animal must consume energy (food) much more frequently than a low metabolism animal.

METONIC CYCLE: The nineteen-year cycle where a correspondence between the day of the year and the lunar phases repeats. For example, the phases of the Moon during the year 2001 will occur on the same day of the year as the phases of the Moon in the year 2020.

MOLECULAR CLOUD: *Gas clouds* that are cool enough for larger *molecules* to form. Since these clouds tend to be very dense, they are the most likely place for the onset of star formation.

MOLECULE: A chemical combination of *elements* that normally has very different properties from its constituent parts. For example, sodium and chlorine will kill you in a high enough dose. Together, as the molecule "sodium chloride," they become ordinary table salt.

MOLTEN: In geology, it is often used to describe melted rock. More generally, however, it is used to describe any thick liquid that is normally a solid.

MOMENTUM (ANGULAR, LINEAR): Angular and linear momentum each have specific formulas. Conceptually, however, we may consider them to represent the tendency for an object to remain rotating (angular) or to remain in motion in a straight line (linear). Momentum is one of the "conserved" quantities in nature where in a closed system momentum remains unchanged.

NEUTRON: A particle in the nucleus of all *atoms* (except for normal hydrogen). It is slightly more massive than the proton and contains no electric charge.

NEUTRON STAR: The tiny remains (less than twenty miles in diameter) of the core of a supernova explosion. It is composed entirely of *neutrons* and is so dense that it is equivalent to cramming two thousand ocean liners in to a *cubic inch* of space.

NORTHERN LIGHTS: See *aurora borealis*.

NUCLEUS: The region of an *atom* that contains *protons* and *neutrons*.

OBLATE SPHEROID: A sphere that is squashed into a shape not unlike a hamburger.

OPEN CLUSTER: See *galactic cluster*.

PARALLAX: The change in position of a star or any other object that appears to occur just because your point of view has shifted. For example, your thumb held at arm's length will appear to be aligned with a different part of the background when you look with your left eye, and then with your right eye.

PERIHELION: The point of closest approach to the Sun for any object (planet, comet, etc.) in a noncircular orbit.

PERIOD: Usually, the time for an object in orbit to complete one orbit. The period of Earth is one year.

PHOTON: Massless particle of light energy. Its energy determines the part of the spectrum where it would be detected. High-energy photons are gamma rays, medium-energy photons are visible light, low energy photons are radio waves.

PLANE: A conceptual region of space that is broad and flat. It is commonly used as a reference to orbits and orientations. For example, "Earth's axis is tilted 23½ degrees from the *plane* of the solar system."

PLASMA: An extremely hot gas (like a normal star) where most of the gas atoms' outer electrons have been stripped away, leaving a charge-filled cloud that responds to magnetic fields. It is sometimes called the "fourth state" of matter.

POLARIS: Also known as the North Star. On the *celestial sphere* it is within one degree of the position that is directly above the Earth's north pole. It is not especially bright, but can be noticed easily due north with an altitude above the horizon that equals the latitude of your location on Earth.

POTENTIAL ENERGY: The energy content that an object has by virtue of its chemical configuration or its position in space. For example, trinitrotoluene (TNT) has enormous chemical potential energy, and water at the top of a waterfall has enormous gravitational potential energy.

PRECESSION: The change in the direction of a nonspherical object's rotation axis by virtue of external forces.

PRIMORDIAL: Generally refers to the chemical or physical conditions that existed at the formation of the Earth, Sun, galaxy, universe, etc.

PROLATE SPHEROID: A sphere that is squished into a shape not unlike a hot dog or an American football.

PROMINENCES: large streams of *plasma* ejected from the Sun's surface. Some extend three or four times the Earth's diameter. Prominences follow the local *magnetic field* lines as they arc upwards and return to the solar surface. They are more common during *sunspot* activity.

PROPER MOTION: The systematic motion of an object as seen against a more distant background. Unlike *parallax*, proper motion is due to the object's actual motion in space, rather than a change in the observer's point of view.

PROTON: The positively charged particle found in the *nucleus* of every *atom*. The number of protons in a *nucleus* defines the identity of an atom. For example, the *element* that has one proton is hydrogen. The element that has two protons is helium. The element that has ninety-two protons is uranium.

QUATERNARY PERIOD: The designation given to the geologic time that is characterized by the appearance and development of the human species. It includes the Pleistocene and Holocene epochs.

RADIATION: Any form of light—visible, infrared, radio, etc. In this nuclear age, however, it has come to mean any particle or form of light that is bad for your health.

RAREFIED: Thin and wispy. Used almost exclusively to describe vacuous gases.

RED SHIFT: The lengthening of the measured wavelength of light due to the motion of the light-emitting object away from you. Since motion is relative, this shift also occurs if you are in motion away from the object.

REDSHIFT (ONE WORD): The general term used to describe the red shifted spectra of nearly all galaxies in the universe. This universal redshift is evidence for our expanding universe.

REFLECTOR TELESCOPE: The form of telescope that uses focusing concave mirrors to collect light. It resembles the "magnifying" flip side of a beauty mirror—except the telescope mirror is a bit more costly.

REFRACTOR: The form of telescope that uses focusing convex lenses to collect light.

RELATIVITY: The general term used to describe Einstein's *Special Theory of Relativity* and *General Theory of Relativity*.

RESOLUTION: The ability of a light-collecting device, camera, telescope, microscope, etc. to convey detail. Resolution is always improved with larger lenses or mirrors.

REVOLUTION: The motion around another object. For example, the Earth revolves around the Sun. It is often confused with *rotation*.

RIGHT ASCENSION: The counterpart of Earth's *longitude* lines onto the celestial sphere. Lines of *longitude* also help determine time zones on Earth. From an astronomer's perspective it is a great advantage to use *time* for right ascension instead of degrees because it then becomes a trivial calculation to determine *when* a star or any other object is visible during the night and during the year. The 360 degrees around the celestial sphere are therefore divided into twenty-four hours of right ascension.

ROTATION: The spinning of an object on its own axis. For example, the Earth rotates once every 23 hours and 56 minutes.

SHOCK WAVE: A concentrated, moving region of sound energy that forms when an object moves faster than the speed of sound in a medium. The crack of a whip, sonic booms, and bomb blasts are all shock waves.

SOUTHERN LIGHTS: See *aurora australis.*

SPACE-TIME: The mathematical combination of space and time that treats time as a coordinate with all the rights and privileges accorded space. It has been shown through the Special Theory of Relativity that nature is most accurately described using a space-time formalism. It simply requires that all events are specified with space *and* time coordinates. The appropriate mathematics does not concern itself with the difference.

SPECIAL THEORY OF RELATIVITY: First proposed in 1905

by Albert Einstein. It provides a renewed understanding of space, time, and motion. The theory is based on two "Principles of Relativity": 1) The speed of light is constant for everyone no matter how you choose to measure it; and 2) the laws of physics are the same in every frame of reference that is either stationary or moving with constant velocity. The theory was later extended to include accelerating frames of reference in the *General Theory of Relativity*. It turns out that the two Principles of Relativity that Einstein *assumed* have been shown to be valid in every experiment ever performed. Einstein extended the relativity principles to their logical conclusions and predicted an array of unusual concepts that include the following:

- There is no such thing as absolute simultaneous events. What is simultaneous for one observer may have been separated in time for another observer.
- The faster you travel the slower your time progresses relative to someone observing you.
- The faster you travel the more massive you become so that the engines of your spaceship are less and less effective in increasing your speed.
- The faster you travel the shorter your space ship becomes—everything gets shorter in the direction of motion.
- *At* the speed of light time stops, you have zero length, and your mass is infinite. Upon realizing the absurdity of this limiting case Einstein concluded that you cannot reach the speed of light.

Experiments invented to test Einstein's theories have verified all of the above predictions precisely. An excellent example is provided by particles that have decay "half-lives." After a predictable time, half of these particles are expected to decay into another particle. When these particles are sent to speeds near the speed of light (in particle accelerators) the half-life increases in the exact amount predicted by Einstein. They also get harder to accelerate which implies that their effective mass has increased.

SPECTRAL TYPE: Any one of several lettered designations that indicate the temperature of a star. In order from hottest to coolest the designations are: O, B, A, F, G, K, M. Historically, stars were classified solely according to features in their spectra. Letters were assigned, in order through the alphabet, to classes of stars. Later, this method proved to be less useful than a classification scheme based on temperature. Many stellar classes were dropped and some were joined with others. What remains is a hodgepodge letter sequence that is the darling of mnemonic writers.

SPECTRUM: The appearance of light after it is separated into its component *wavelengths*. Visually, the human eye detects *wavelengths* by colors.

SPHERE: The only solid shape where every point on its surface is the same distance from its center.

SUNSPOTS: Small, circular regions of the Sun's surface that are somewhat cooler than the surrounding areas. This temperature contrast makes sunspots appear dark against their brighter background. They move with the Sun's surface and tend to avoid the polar and equatorial regions. Sunspots commonly travel in pairs due to their association with the Sun's *magnetic field*. They come and go in "waves" that define the eleven-year solar activity cycle. The average sunspot is about two or three times larger than the Earth.

TELESCOPE (GAMMA, X-RAY, ULTRAVIOLET, OPTICAL or VISIBLE, INFRARED, MICROWAVE, RADIO): Astronomers have designed special telescopes and detectors for each part of the *spectrum*. Some parts of this spectrum do not reach Earth's surface. To see the gamma rays, X rays, ultraviolet light, and infrared light that is emitted by many cosmic objects, these telescopes must be lifted into orbit above the absorbing layers of Earth's atmosphere. While all the telescopes are of different design, they do share three basic principles: 1) They collect photons; 2) they focus photons; and 3) they record the photons with some sort of detector.

THERMAL ENERGY: The energy contained in an object (solid, liquid, or gaseous) by virtue of its atomic or molecular vibrations. The *kinetic energy* of these vibrations is the official definition of temperature.

THERMONUCLEAR: Any process that pertains to the behavior of the atomic *nucleus* in the presence of high temperatures.

TIME ZONE: Any one of twenty-four equal divisions of Earth's *longitude* that are all separated by exactly one hour. Each zone is assigned the same time for the convenience of travel and commerce. In practice, time zone boundaries tend to follow state and national borders.

TWENTY-ONE-CENTIMETER LINE (21 CM): A feature in the *spectrum* of hydrogen gas. The lone electron of the hydrogen atom occasionally flips its spin orientation in the atom. This emits a long 21cm radio wave *photon* that easily penetrates the obscuring clouds and dust of the interstellar medium.

VARIABLE STAR: A star that for any reason varies its luminosity by a substantial amount. Some stars vary their luminosity episodically. They are called cataclysmic variables. Some stars vary their luminosity because they are in a tight orbit with another star that periodically blocks their light. These are called eclipsing variables. Other stars vary their luminosity intrinsically. They are considered ordinary variable stars.

VEGA: The seventh brightest star in the sky. The brightest star in the constellation Lyra. Vega is one of a handful of stars where "debris" (possibly planets or planetoids) was detected in orbit around it.

WAVELENGTH: The length of a repeating component of a wave. It is a very useful term that applies to sound, light, trucks in a convoy, ripples on water's surface, etc.

SUGGESTED BIBLIOGRAPHY

Magazines

If you get excited about occultations, eclipses, telescope conventions, etc., then try:

Sky and Telescope
Sky Publishing Company
49 Bay State Road
Cambridge, Mass. 02238

If you like pretty pictures of astronomical objects and colorful page layouts then try:

Astronomy
P. O. Box 1612
Waukesha, Wisc. 53187-9950

Both *Sky and Telescope* and *Astronomy* magazines are published monthly and provide star charts, extensive telescope advertisements, news events, and classified ads. They are indespensible to the amateur astronomer.

If you are a "weekend" astronomer and if you consider yourself a lay reader, and if you would like to communicate with Merlin then try:

Star Date
McDonald Observatory
The University of Texas at Austin
Austin, Tex. 78712

Star Date, published bimonthly, provides star charts and a sky calendar. The magazine will not only provide diverse reading, it will also look good on your coffee table.

Another monthly magazine for the lay reader is:

Griffith Observer
2800 East Observatory Road
Los Angeles, Calif. 90027

In addition to its star charts and monthly sky calendar, it often features diverse articles in subareas of astronomy that include astroarchaeology, and the history of astronomy.

If you take science very seriously then there is no substitute for:

Scientific American
415 Madison Avenue
New York, N.Y. 10017

In spite of what they may claim it is not a magazine for the lay reader.

If you like up-to-the-minute discoveries and developments in all areas of science then try:

Science News
1719 N Street, N.W.
Washington, D.C. 20036

It is written by science writers for the lay reader and is published weekly.

Books

The First Three Minutes. S. Weinberg. New York: Bantam, 1979.

The Left Hand of Creation. J. D. Barrow and J. Silk. New York: Basic Books, 1983.

Readable accounts of the origin of the universe described in full detail.

Cosmos, C. Sagan. New York: Random House, 1981.

The companion volume to the popular PBS series of the same name.

The ABC of Relativity. B. Russell. New York: Mentor Books, 1985.

Time, Space, and Things. B. K. Ridley. New York: Cambridge University Press, 1986.

The titles say it all.

Einstein for Beginners. J. Schwartz and M. McGuinness. New York: Pantheon Books, 1979.

Was Einstein Right? Putting General Relativity to the Test. C. M. Will. New York: Basic Books, 1979.
If you can't understand what Einstein said then these two accounts are your next best description of gravity.

Build Your Own Telescope. R. Berry. New York: Scribner, 1985.
Astrophotography: A Step by Step Approach. R. T. Little. New York: Macmillan, 1985.
Astrophotography for the Amateur. M. A. Covington. New York: Cambridge University Press, 1985.
If you like "hands-on" astronomy, then the above books are for you.

Norton's Star Atlas. P. Norton. Cambridge, Mass.: Sky Publishing, 1978.
Whitney's Star Finder. C. Whitney. New York: Knopf, 1977.
The time-honored standards for finding stars at night.

Galaxies. T. Ferris. San Francisco: Sierra Club Books, 1981.
Man Discovers Galaxies. R. Berendzen, R. Hart, and D. Seely. New York: Columbia University Press, 1984.
Accept no substitutes. Two indispensable books on Merlin's favorite topic.

Sleepwalkers. A. Koestler. New York: Grosset & Dunlap, 1963.
The delightful account of some famous astronomers of the past.

Newton's Philosophy of Nature. H. S. Thayer. New York: Hafner Press, 1974.

It sifts through Newton's writings and brings you the "Best of. . . ." It includes some of Newton's letters and annotated excerpts from *Principia* and *Optiks*.

The New Solar System. ed. J. Kelly Beatty and Andrew Chaikin. Cambridge, Mass.: Sky Publishing, 1990.

The Planets. Scientific American Books. San Francisco: W. H. Freeman, 1983.

Two comprehensive accounts of all members of the solar system.

Black Holes: The Edge of Space, The Edge of Time. W. Sullivan. New York: Anchor Press/Doubleday, 1979.

Black Holes, Quasars, and the Universe. H. L. Shipman. Boston: Houghton-Mifflin, 1980.

Geometry, Relativity, and the Fourth Dimension. R. Rucker, New York: Dover, 1977.

Serious but readable treatments of astronomy exotica.

The Search for Extraterrestrial Intelligence, NASA. Washington, D.C.: GPO, 1977.

Intelligent Life in the Universe. C. Sagan and I. S. Shklovskii. San Francisco: Holden-Day, 1966.

What extraterrestrial life might be like and how to go about searching for it.

ABOUT THE AUTHOR

*N*EIL DE GRASSE TYSON, 38, was born and raised in New York City. He attended the Bronx High School of Science, earned his B.A. in Physics from Harvard and his Ph.D. in astrophysics from Columbia. He recently joined Stephen Jay Gould as *Natural History*'s only other monthly essayist with his column "Universe." Tyson currently holds a joint position as research scientist at Princeton and as the Frederick P. Rose Director of the Hayden Planetarium, where he is overseeing the $40 million rebuilding of one of the nation's greatest astronomical attractions.